Land and Freshwater Snails of Tahiti and the other Society Islands

Justin Gerlach

© Justin Gerlach
Phelsuma Press, Cambridge

ISBN: 978-1-6781-9674-5

Land and Freshwater Snails of Tahiti and the other Society Islands

The snail fauna of French Polynesia's Society Islands is best known for the highly diverse *Partula* species, but comprises a total of 166 species (of which the Partulidae is the largest family).

Although a few snails records date back to the 18th century most derive from the collections made by Andrew Garrett in the 1860-80s and the Mangareva Expedition of 1934. A few more species have been recognised since then, but most of these are recent introductions. This book covers all the recorded land and freshwater snails, excluding littoral species such as the ellobiid *Melampus* species (*M. caffer, M. fasciata, M. luteus, M. monile, M. philippi* and *M. striatus*). Higher-level classification follows Bouchet *et al*. (2005).

Tahiti is the most diverse island (126 species), with its large area and diverse habitats. At the other extreme small Maupiti and Mehetia have only 16 and 9 species respectively. There are no data from the coral islands of Tetiaroa, Maiao, Tupai, Maupihaa, Maunuae and Motu One, but these are expected to support only a few species of cosmopolitan or introduced species.

In the following text the following abbreviations are used: ANSP – Academy of Natural Sciences, Philadelphia; BPBM – Bernice P. Bishop Museum, Hawaii; MNHN – Museum national d'Histoire naturelle, Paris; NHMUK – Natural History Museum, London; USNM – Smithsonian National Museum of Natural History, Washington. Photographs of specimens not credited to an institution are in the author's collection.

All species are shown with a scale bar in millimetres. Figures of some species are colour coded with their island of origin, as in the map below.

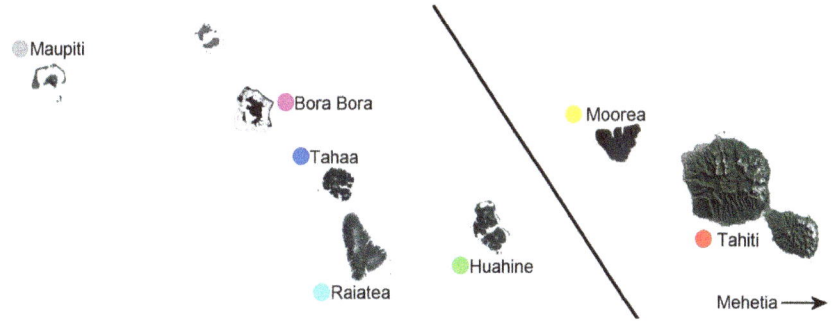

Family	endemic	indigenous	introduced	Mehetia	Tahiti	Moorea	Huahine	Raiatea	Tahaa	Bora Bora	Maupiti
Helicinidae	14	5	0		12	5	2	7	3	6	
Hydrocenidae	0	3	0	1	3	3	3	3	1	1	1
Neritidae	0	8	0		8	8	3	3			
Thiariidae	0	1	2		3	2	1	2	1		
Assimineidae	13	3	0	1	7	7	4	4	2	2	1
Truncatellidae	1	1	0		2		1				
Lymnaeidae	0	0	3		3						
Physidae	0	0	1		1	1		1			
Succineidae	12	2	0		12	3	1	1	1		
Partulidae	56	0	0		15	9	4	24	8	2	1
Achatinellidae	1	6	0	2	5	5	4	3	3	3	3
Valloniidae	0	1	0			1		1			
Vertiginidae	4	3	1	2	7	2	3	2	1	2	1
Achatinidae	0	0	1		1	1	1	1	1	1	1
Subulinidae	0	0	8	1	6	7	4	6	3	3	2
Spiraxidae	0	0	1		1	1	1	1	1	1	1
Streptaxidae	0	0	1		1	1	1	1			1
Charopidae	6	0	0			1	1	1		1	
Endodontidae	32	0	0		13	6	5	4	2	5	2
Euconulidae	22	4	0	2	15	8	6	7	2	3	
Trochomorphidae	5	0	0		2	1	1	2	1		
Zonitidae	0	0	0		1					1	
Helicarionidae	0	0	1		1	1	1	1	1	1	1
Limacidae	0	0	1		1			1			
Philomycidae	0	0	1			1					
Braybaenidae	0	0	1		1	1	1	1			1
Veronicellidae	0	0	3		2	2			1	1	
TOTAL	166	37	26	9	126	78	52	77	32	32	16

Cycloneritimorpha
Superfamily Helicinoidea
Helicinidae

Three genera have been recorded: *Aphanoconia* (lenticular), *Orobophana* (discoidal to hemispherical, distinctly robust), *Nesoicina* (lenticular), *Pleuropoma* (roundedly lenticular) and *Sturanya* (globosely discoidal). These generic assignments are probably unreliable as *Aphanoconia* is an invalid name, and most species are often placed in *Pleuropoma*, but that genus is restricted to the Philippines, and *Sturanya* is from the western Pacific.

In addition to the relatively well defined species there are a number of records that appear to be misidentified or mislabeled: *Pleuropoma pictum* 'Pease, 1865' (nomen nudum, from Tahiti ANSP 14447), *P. tectiformis* Mousson ('Tahiti' ANSP 362523). *Orobophana uberta* (Gould, 1847) types were originally labeled 'Taheiti' in error (Johnson 1964); the species is restricted to the Hawaii islands. This appears to be the same as *Helicina constricta* Pfeiffer, 1846-53 also from 'Otaheite' (Neal 1934) (photo NHMUK - Cuming specimen).

Aphanoconia corrugata (Pease, 1865)
Helicina corrugata Pease, 1865
Aphanoconia corrugata (Pease) Wagner 1907-11
Raiatea, Tahaa – historically scarce, on ground or low vegetation, now extinct. Photo NHMUK.

Aphanoconia exigua (Hombron & Jacquinot, 1848)
Helicina exigua Hombron & Jacquinot, 1848
Helicina inconspicua Pfeiffer, 1848
Helicina decolorata Schmeltz 1874 nomen nudum
Orobophana inconspicua (Pfeiffer) Wagner 1905
Aphanoconia (Sphaeroconia) inconspicua (Pfeiffer) Wagner 1907
Gambiers - Tahiti, Huahine, Maupiti. On ground from coast to 300m. Extinct? NHMUK 20030002 labeled *Helicina parvula* Pease is this species. Photo NHMUK (*H. inconspicua*).

Aphanoconia faba (Garrett, 1884)
Helicina faba 'Pease' Garrett 1884
Aphanoconia discoidea Wagner, 1905 (*faba* and *tumidior*).
Moorea, Raiatea. On vegetation. Extinct? Photo syntype ANSP 14544.

Aphanoconia kusteriana (Pfeiffer, 1848)
Sulfurina Kusteriana Pfeiffer, 1848
Tahiti, Bora Bora. Extinct? Figure from Wagner (1905).

Aphanoconia minuta Sowerby, 1842
Helicina minuta Sowerby, 1842

Pleuropoma (Aphanoconia) minutum Sowerby, 1842
Orobophana minuta (Sowerby) Wagner 1905
Aphanoconia (Sphaeroconia) minuta (Sowerby) Wagner 1907-11
Australs - Tahiti, Moorea (in litter). Extinct? Figure from Wagner (1905).

Aphanoconia pentheri Wagner, 1904
Aphanoconia pentheri Wagner, 1904
Tahiti. Figure from Wagner (1905).

Aphanoconia rustica (Pfeiffer, 1852)
Helicina pallida Pfeiffer, 1848
Helicina rustica Pfeiffer, 1852
Helicina rugulosa Pease, 1868
Sturanya rustica (Pfeiffer) Wagner 1905
Widespread in Societies, specific records from Tahiti (coastal to 150m), Bora Bora. Extinct? Smaller than *rusticana*. Photo NHMUK.

Aphanoconia rusticana Wagner, 1907
Aphanoconia (Sphaeroconia) rusticana Wagner 1907-11
Bora Bora. Extinct. Figure from Wagner (1905).

Nesiocina discoidea (Pease, 1867)
Helicina discoidea Pease, 1867
Aphanoconia discoidea (Pease) Wagner 1905
Nesiocina discoidea (Pease) Richling & Bouchet 2013
Tahiti, Raiatea, Tahaa – historically present on ground in lowland forest, now extinct on Raiatea and Tahaa, restricted to higher altitudes on Tahiti. Photo lectotype ANSP 14398.

Nesiocina solidula (Sowerby, 1839)
Helicina solidula Sowerby, 1839
Helicina colorata Pease, 1868
Orobophana colorata (Pease) Wagner 1905, including ssp. *raiateae* Wagner and *solidula* (Gray)
Orobophana solidula (Sowerby) Wagner, 1907, including spp. '*raiatteae*' Wagner and *colorata* (Pease)
Nesiocina solidula (Sowerby) Richling & Bouchet 2013

Nesiocina discoidea

Australs. Tahiti (*colorata*), Raiatea (*raiateae*). Extinct on Raiatea, possibly extinct on Tahiti. Photo NHMUK.

Orobophana albolabris Hombron & Jacquinot, 1848
Helicina albolabris Hombron & Jacquinot, 1848
Helicina solida Pease, 1865
Orobophana albolabris (Hombron & Jacquinot) Wagner 1907-11
Tahiti (widespread in 1928). On trees and bushes. Extinct. Specimens of '*Helicina solida*' from Raiatea (NHMUK) are errors for *Nesiocina solidula*. Photo NHMUK. Raiatea (left) and '*Helicina crassilabris* Philippi, 1847' (misidentified) (right).

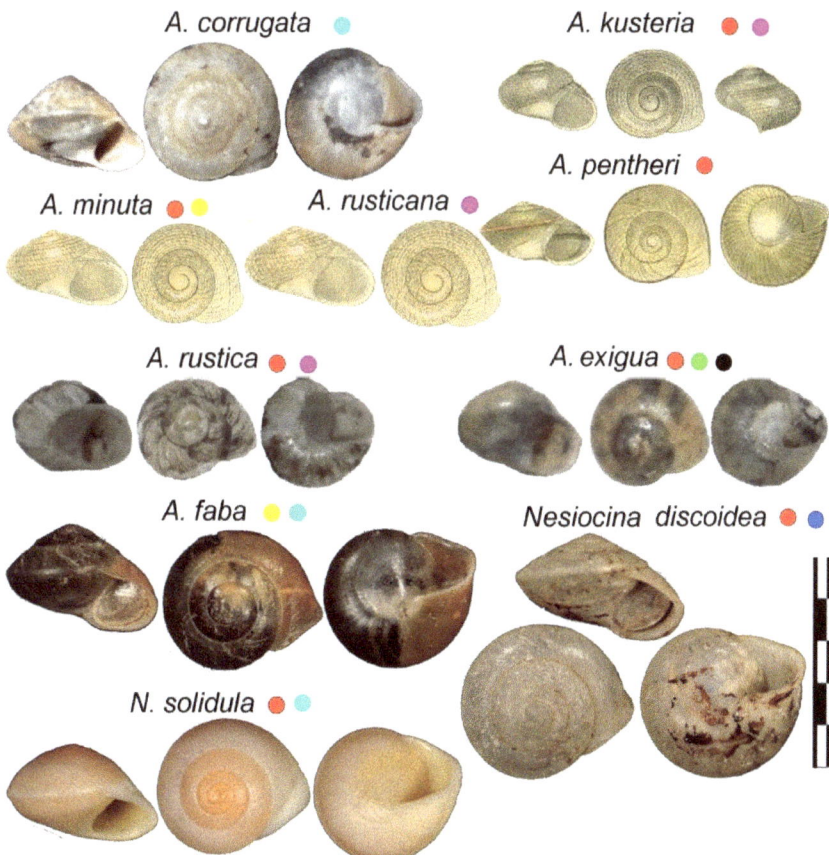

Orobophana maugeriae (Gray, 1825)
Helicina maugeriae var. *rubicunda* Pease, 1868
Helicina maugeriae Gray, 1825
Orobophana maugeriae (Gray) Wagner 1905
Helicina rubicunda 'Pease' Carpenter 1864
Helicina maugeriae var. *albinea* Pease, 1871
Helicina bella 'Pease' Schmeltz 1877
Orobophana maugeriae albinea (Pease) Wagner 1905
Raiatea (on foliage in upper parts of valley on east and west, *rubicunda* from ground in lower parts of Viaau, Fatimu), Tahaa (*albinea*: valley in east and Hamene). Extinct. Photo ANSP 14512 (*rubicunda* lectotype).

Orobophana miniata (Lesson, 1831)
Helicina miniata Lesson, 1831
Helicina Rolvii Carpenter 1864 nomen nudum
Helicina rufescens Pease 1865, nomen nudum
Orobophana miniata (Lesson) Wagner 1907-11
Bora Bora, at the foot of trees and on vegetation on the mountain. Extinct. '*Helicina rufescens*' specimens are doubtfully from Tahiti (ANSP 14513, no data) and Raiatea (BPBM). Photo syntype MNHN IM-2000-24801.

Orobophana pacifica (Pease, 1865)
Helicina pisum Hombron & Jacquinot, 1848 non Phillipi
Helicina pacifica Pease, 1865
Helicina straminea Pease 1865 nomen nudum
Helicina tahitiensis Pease, 1871
Helicina flavescens Pease Garrett 1884
Orobophana tahitensis (Pease) Wagner 1905
Cooks - Tahiti, Huahine, Raiatea, Bora Bora. Abundant in lowlands, close to shore, under stones. Not located recently except for very old shells on Bora Bora. Photos *pisum* type MNHN (left, apertural only) and NHMUK *straminea* Tahiti (centre and right).

Pleuropoma raiatensis (Garrett, 1884)
Helicina raiatensis Garrett, 1884
Pleuropoma (Orobophana) solidula Gray, 1825
Raiatea (west). In leaf litter. Probably extinct. Photo lectotype ANSP 14455.

Pleuropoma simulans (Garrett, 1884)
Helicina simulans Garrett, 1884
Tahiti. On bushes. Not located recently. Photo lectotype ANSP 14458.

Pleuropoma subrufa Garrett, 1884
   Helicina subrufa 'Pease' Garrett, 1884
   Raiatea, Bora Bora. On ground. Extinct. Photo lectotype ANSP 14463.
Sturanya multicolor Gould, 1847
   Helicina multicolor Gould, 1847
   Sturanya multicolor (Gould) Wagner 1905
   West Pacific - Tahiti, Moorea. Last recorded 1852. Photo lectotype USNM 5512.

Superfamily Hydrocenoidea
Hydrocenidae

Three minute *Georissa* species differ in size and sculpture, from the smallest and most sculptured (*G. striata* 1.7mm), larger and sculptured (*G. insularis* 1.7-2mm) to the largest and smoothest (*G. parva* 2.7mm).

Georissa (Chondrella) insularis (Crosse, 1865)
   Hydrocena insularis Crosse, 1865
   Omphalotropis insularis Pease 1865
   Atropis insularis Pease 1871
   Chondrella insularis Crosse Garrett 1884
   Realia insularis Pfeiffer 1858
   Gambiers – Tahiti, Moorea, Huahine, Raiatea, Tahaa, Maupiti.
Georissa (Chondrella) parva (Pease, 1865)
   Cyclostoma parvum Pease, 1865
   Georissa parva (Pease) Garrett 1887
   Chondrella parva (Pease) Garrett 1884
   Marquesas - widespread including Tahiti, Moorea, Huahine, Raiatea, Bora Bora. On ground.
Georissa (Chondrella) striata (Pease, 1871)
   Chondrella striata Pease, 1871
   Australs - Mehetia, Tahiti, Moorea, Huahine, Raiatea.

G. insularis  G. parva  G. striata

Superfamily Neritoidea
Neritidae
In addition to these largely freshwater species, estuarine areas may be occupied by mainly coastal species such as *Nerita (Melanerita) melanotragus* Smith, 1884, *Neritina (Vittoida) variegata* (Lesson, 1831).

Clithon (Clithon) diadema (Recluz, 1841) ***recluziana*** (Guillou, 1841)
Nerita recluziana Guillou, 1841
Clithon (Clithon) diadema (Recluz) Starmuhlner 1976
Clithon diadema recluziana (Guillou) Cowie 1998
Pacific – Tahiti, Moorea. Photos MNHN *C. recluziana* type.

Clithon spinosum (Sowerby, 1825)
Clithon undatus Lesson, 1831
Neritina spinosa Martens 1879
Clithon (Clithon) spinosus (Budgin, in Sowerby) Starmuhlner 1976
Clithon spinosa (Budgin) Resch, Barnes & Craig 1990
Pacific - Tahiti, Moorea, Huahine, Raiatea. Estuarine. Pointer & Marquet (1990) considered this to be the only *Clithon* species in French Polynesia. Museum specimens of *C. corona* (Linnaeus, 1758) from Tahiti and Raiatea appear to be mislabeled or misidentified. Photo *C. undatus* type MNHN.

Neritepteron (Neripteron) taitensis (Lesson, 1831)
Neritina taitensis Lesson, 1831
Neritina dilatata Broderip, 1832
Neritina nux Broderip, 1833
Neritina chlorostoma Sowerby, 1833
Neritina florida Recluz, 1850
Neritina auriculata (Lamarck) f. *tahitensis* Starmuhlner 1976
Neripteron (Neripteron) taitensis (Lesson) Hayes 2001
Samoa to Marquesas – Tahiti (Pt Venus), Raiatea, Huahine, Moorea. Lower courses of rivers with tidal influence. Specimens labeled *Neripteron bicanaliculatum* (Recluz, 1843) from 'Tahiti' are probably mislabeled or misidentified. Photo *S. taitensis* and *S. florida* types MNHN.

Neritilia rubida (Pease, 1865)
Neritina rubida Pease, 1865
Neritilia rubida (Pease) Starmuhlner 1976
Indo-Pacific – Tahiti, Moorea. Fast flowing streams to 1km from sea.

Neritina (Neritina) canalis Sowerby, 1825
   Neritina canalis Sowerby Martens 1879
   Neritina (Neritina) canalis Sowerby Starmuhlner 1976
   Pacific – Tahiti, Moorea, Huahine, Raiatea. Recorded from upper parts of rivers. There is some confusion with *N. pulligera* Linnaeus, 1767 and Benthem-Jutting (1963) considered the two taxa to be synonymous. Although Pilsbry listed *N. pulligera* Linnaeus, 1767 var. *ovalis* from Tahiti, Starmuhlner (1976) considered *N. pulligera* to be an Indo-west Pacific species. Specimens labeled *N. porcata* Gould, 1847 may be this species as well.

Neritina (Neritina) unidentata Recluz, 1850
   Neritina (Neritina) unidentata Recluz, 1850
   A species of uncertain status, recorded from Tahiti by Recluz (1850) but not recognised subsequently. It may be a juvenile of *C. spinosus*. Photo type MNHN.

Neritina (Vittina) turrita (Gmelin, 1790)
   Neritina (Vittina) turrita (Gmelin) Starmuhlner 1976
   Pacific - Tahiti, Moorea, Raiatea. Common in lowland streams, including roadside ditches connecting to estuaries.

Platynerita rufa Kano & Kase, 2003
   Neritina rubida Pease, 1965 (part)
   Platynerita rufa Kano & Kase, 2003
   Pacific – Tahiti, Moorea (Vaioro river at Afareaitu falls). Fast streams.

Septaria taitana Mousson, 1869
   Septaria taitana Mousson, 1869
   Septaria borbonica depressa Reich, 1937
   Septaria macrocephala (Guillou in Recluz) Starmuhlner 1976
   Septaria porcellana (Linne) Starmuhlner 1976
   Septaria procellana f. *depressa* Starmuhlner 1976
   Pacific - Tahiti (Papenoo steam), Moorea, Raiatea. The only Society Islands *Septaria* species (Hayes 2001). *Septaria lineata* (Lamarck, 1816) specimens from 'Tahiti' are probably misidentifications or mislabeling.

Superfamily Cerithioidea
Thiaridae
Melanoides tuberculata (Muller, 1774)
   Melanoides corporosa Gould, 1847
   Melania gracilina Gould, 1859

Thiara scopulus Reeve, 1860
Melania luteola Dunker, 1866
Melania Tahitensis Dunker 1866
Melania minuta Tryon, 1866
Melania unicolor Tryon Horst & Schepman 1908
Stenomelania lancea Lea Horst & Schepman 1908
Melanoides minuta Tryon Baker 1964
Melanoides (Melanoides) tuberculatus (Muller) Starmuhlner 1984
Pantropical (introduced – Cowie 1998) – Tahiti, Moorea, Huahine, Raiatea, Tahaa, common on all islands. Several names have been applies to museum specimens from French Polynesia which are probably misidentifications of *M. tuberculata*: *M. indefinita* Lea & Lea, 1851 (as *M. newcombi* Lea, 1856 and *Melania contigua* Pease, *Thiaria cybele* (Gould, 1847)), *T. mitra* Meuschen, 1863. Photo USNM 5561 (*M. corporosa* holotype), ANSP 26525 ('*Melania unicolor*' Pease) ANSP 26375 (*M. minuta* lectotype).

Stenomelania persulcata Mousson, 1869
Melanoides (Stenomelania) arthurii (Brot, 1870) Starmuhlner 1976
Melanoides persulcata Mousson Cowie 1998
Pacific – Tahiti, Raiatea. Other names have applied to museum specimens of *Stenomelania* from French Polynesia are probably misidentifications: *S. plutonia* Hinds, 1844, *S. scipio* (Gould, 1847). Smoother than *Melanoides tuberculata*.

Tarebia granifera (Lamarck, 1816)
Tarebia tahitensis Brot, 1877 nomen nudum
Thiara granifera (Lamarck) Starmuhlner 1976
Introduced? - Tahiti, Moorea. First recorded in 1877 (Tahiti).

Other freshwater species

These pulmonate snails are placed here for convenience, occurring in the same habitats as the nerites.

Superfamily Lymnaeoidea
Lymnaeidae

Lymnaea (Radix) sp.

Lymnaea (Radix) sp. Starmuhlner 1976
Introduced - Tahiti (Papeete). First recorded in 1975. Figure from Starmuhlner 1976.

Superfamily Planorboidea
Planorbidae
    Pettancylus* cf. *noumeensis (Crosse, 1871)
        Ferrissia (P) sp. Starmuhlner 1976
        Introduced – Tahiti (Orofara). Identification uncertain (suggested by Haynes, 1984, but Marquet 1993 listed it as unidentified). First recorded in 1975. Figure from Starmuhlner 1976.
    Indoplanorbis exustus (Deshayes, 1834)
        Indoplanorbis exustus (Deshayes) Starmuhlner 1976
        Introduced – Tahiti (Papeete). First recorded in 1975.

Physidae
    Physastra nasuta Morlet, 1857
        Physastra tahitensis Clessin 1886
        Physa sp. Starmuhlner 1976
        Glyptophysa (Physastra) sp. Walker 1988
        Physastra nasuta (More) Cowie 1998
        Introduced – Tahiti, Moorea, Raiatea. First recorded on Tahiti,1800s.

Superfamily Rissoidea
Assimineidae
    Assimineids are mostly marine snails but three genera include terrestrial species: some *Assiminea* are coastal, whereas *Garrettia* and *Omphalotropis* and related genera are fully terrestrial. In addition to the terrestrial species two intertidal or littoral species have been recorded from the Society Islands (*Taiwanassiminea affinis* (Boettger, 1887) [=*Assiminea affinis*] from Raiatea and *Cyclomorpha flava* (Broderip, 1832)).
    Some records are probably mislabeled specimens: *Omphalotropis variabilis* (Pease, 1865) (ANSP 83197) and *O. subimperforata* Boettger, 1916 (ANSP 7875, 83196) are from 'Huahine or Ruratonga'. Others may be mislabeled (*O. bulimoides* (Hombron & Jacquinot, 1858) ANSP 13302, 'Tahiti' Pease specimen, may be an error for *O. ventricosa* (Hombron & Jacquinot, 1848 [=*Cyclostoma ventricosa, Realia ventricosa* Pfeiffer 1854, *Atropis ventricosa* Martens 1866] which Garrett (1884) thought could be *O. viridescens*. 7x5mm it is larger than recorded Tahiti species.
    Assiminea parvula (Mousson, 1865)
        Hydrocena nitida Pease, 1865
        Assiminaea (Hydrocena) nitida Pease 1869
        Indo-Pacific – Mehetia, Tahiti, Moorea, Huahine, Raiatea. Coastal.

Eastern species

Garrettia biangulata (Pease, 1864)
> *Cyclostoma biangulata* Pease, 1864
> *Diadema biangulata* (Pease) Garrett 1884
> *Garrettia Scalariformis* 'Pease' Paetel 1873
> *Garrettia biangulata* (Pease) Cooke & Clench 1943
> Cooks – Moorea (under leaves in southwest). Garrett (1884) thought it introduced and this is the only record of *Garrettia* outside of the Cook islands. Figure from Garrett 1884.

Atropis abbreviata (Pease, 1864)
> *Realia abbreviata* Pease, 1864
> *Omphalotropis abbreviata* (Pease) Pease 1869
> *Atropis abbreviata* (Pease) Pease 1871
> Tahiti (northwest side), on ground. Figure from .Pease 1869.

Atropis bythinellaeformis Garrett, 1884
> *Atropis bythinellaeformis* Garrett, 1884
> Tahiti (one valley in north at 500m), Moorea (north coast, lowland). Photo lectotype ANSP 14251.

Atropis obesa (Garrett, 1884)
> *Atropis obesa* Garrett, 1884
> Tahiti (one valley in northwest). Photo lectotype ANSP 14234.

Atropis vescoi Dohrn, 1859
> *Hydrocena Vescoi* Dohrn, 1859
> *Omphalotropis Vescoi* Pease 1869
> *Atropis Vescoi* (Dohrn) Pease 1871
> *Realia Vescoi* (Dohrn) Martens & Langkavel 1871
> Tahiti (north, side of a ravine at 500m). Photo NHMUK

Omphalotropis scitula (Gould, 1847)
> *Cyclostoma scitulum* Gould, 1847
> *Omphalotropis scitula* (Gould) Pfeiffer 1851
> *Hydocena scherzeri* Pfeiffer & Zelebor, 1867
> *Atropis scitula* (Gould) Pease 1871
> Pacific – Tahiti (northwest), Moorea. Historicaly plentiful on the ground. Photo syntype ANSP 332959.

Omphalotropis oblonga (Pfeiffer, 1855)
> *Hydrocena oblonga* Pfeiffer, 1855
> *Omphalotropis oblonga* (Pfeiffer) Pease 1869
> *Atropis oblonga* Pease 1869
> *Realia oblonga* Pfeiffer 1876

Western species

Moorea (historically abundant in north), on ground. Figure from Garrett 1884.

Omphalotropis terebralis Gould, 1847
Cyclostoma terebrale Gould, 1847
Omphalotropis terebralis (Gould) Pfeiffer 1851
Atropis terebralis (Gould) Pease 1871
Atropis gouldi Garrett MS in Garrett 1884
Tahiti (always rare, northwest), Moorea (historically plentiful in one area), Bora Bora (historically abundant). On ground at 150m). Photo NHMUK (*gouldi*).

Omphalotropis huaheinesis (Pfeiffer, 1855)
Hydrocena Huaheinensis Pfeiffer, 1855
Assiminea Huaheinensis Martens 1866
Omphalotropis Huaheinensis (Pfeiffer) Pease 1869
Realia Huaheinensis Pfeiffer 1876
Omphalotropis robusta Crosse in Pease, 1869
Moorea, Huahine (widespread), Raiatea (3-4 valleys on west side). On ground; still present on Huahine, status on Raiatea uncertain. Photo *O. robusta* type MNHN.

Omphalotropis boraborensis (Dorhn, 1859)
Omphalotropis boraborensis Dorhn, 1859
Atropis boraborensis (Dorhn) Pease 1871
Realia boraborensis (Dorhn) Pfeiffer 1876
Bora Bora. Historically plentiful on ground. Photo NHMUK (as '*O. filosa*').

Omphalotropis costata (Pease, 1868)
Realia (Scalinela) costata Pease, 1868
Scalinella costata (Pease) Pease, 1869
Realia costata (Pease) Pfeiffer 1876
Tahaa, historically common on ground in lowlands. Photo NHMUK.

Omphalotropis moussoni Garrett, 1884
Scalinella Moussoni 'Semper' Garrett, 1884
Maupiti. Historically common on ground. Figure from Garrett 1884.

Omphalotropis producta (Pease, 1865)
Realia producta Pease, 1865
Realia elongata Pease, 1868
Hydrocena raiateensis Mousson, 1869
Omphalotropis producta (Pease) Pease 1869
Omphalotropis elongata (Pease) Pease 1869

Atropis elongata (Pease) Pease 1871
Atropis producta (Pease) Pease 1871
Realia raiateensis (Mousson) Pfeiffer 1876
Raiatea, Tahaa. On ground. Photo syntype ANSP 13353.
Omphalotropis tahitensis (Pease, 1861)
Cyclostoma tahitensis Pease, 1861
Scalinella tahitensis (Pease) Pease 1869
Realia Tahitensis (Pease) Pfeiffer 1876
Huahine (widespread). On ground. Photo syntype ANSP 13366.
Omphalotropis viridescens (Pease, 1861)
Cyclostoma viridescens Pease, 1861
Omphalotropis viridescens (Pease) Pease 1869
Atropis viridescens (Pease) Garrett 1884
Huahine (widespread), Raiatea (one valley in southeast). On ground; still present on Huahine. Photo type MNHN.

Truncatellidae
Taheitia pallida Pease, 1868
Taheitea pallida Pease, 1868
Taheitia pallida Pease Clench & Turner 1948
Indo-Pacific - Widespread in Societies. Abundant in lowlands (Garrett 1884). Photo NHMUK.
Taheitia porrecta Gould, 1847
Truncatella porrecta Gould, 1846
Taheitia porrecta (Gould) Pease 1871
Tahiti (one mile up Papenoo valley). Figure from Clench & Turner (1948).

T. porrecta   *T. pallida*

Pulmonata, Stylommatophora,
Superfamily Succineoidea
Succineidae

There is a significant radiation of Society Islands succinedis. In addition to the following species there are dubious records. Part of the type series of *Succinea venusta* Gould, 1846 is labeled 'Taheiti' but this is an error (Yeung *et al.* 2017) and the species is exclusively Hawaiian. Additionally the 'Tahiti' records of *S. simplex* Pfeiffer are probably misidentified

Succinea amoi Cooke & Clench, 1945
Succinea amoi Cooke & Clench, 1945
Tahiti (slopes of Orofena and Mt. Marau, from above 150m). Figure from Cooke & Clench 1945 and author's specimen.

Succinea bernardi Recluz, 1852
Succinea Bernardi Recluz, 1852
Tahiti (Orofena waterfall – an amphibious species). Figure from Cooke & Clench 1945.

Succinea costulosa Pease, 1865
Succinea costulosa Pease, 1864
Cooks - Tahiti (only in Fautaua valley, on foliage). Figure from Cooke & Clench 1945.

Succinea dolphin Cooke & Clench, 1945
Succinea wallisis dolphin Cooke & Clench, 1945
Succinea dolphin Cooke & Clench Patterson 1989
Tahiti (Tautira, 100m). Figure from Cooke & Clench 1945.

Succinea humerosa Gould, 1846
Succinea humerosa Gould, 1846
Succinea Tahitensis Pease (non Pfeiffer), 1871
Tahiti. Historically rather common, widespread on the ground. Photo NHMUK (Cuming collection).

Succinea infundibuliformis Gould, 1846
Succinea infundibuliformis Gould, 1846
Truella infundibuliformis (Gould) Pease 1871
Tahiti (southwest and on Mt. Marau, above 1100m, on ground and low vegetation). Also on Moorea according to Gould but Garrett could not find it there. Figure from Gould 1846.

Succinea modesta Gould, 1846
Succinea modesta Gould, Gould 1852
Samoa – Tahiti (Mt. Marau), Moorea (Mt. Tohiea). Photo syntype USNM 5424.

Tahiti species

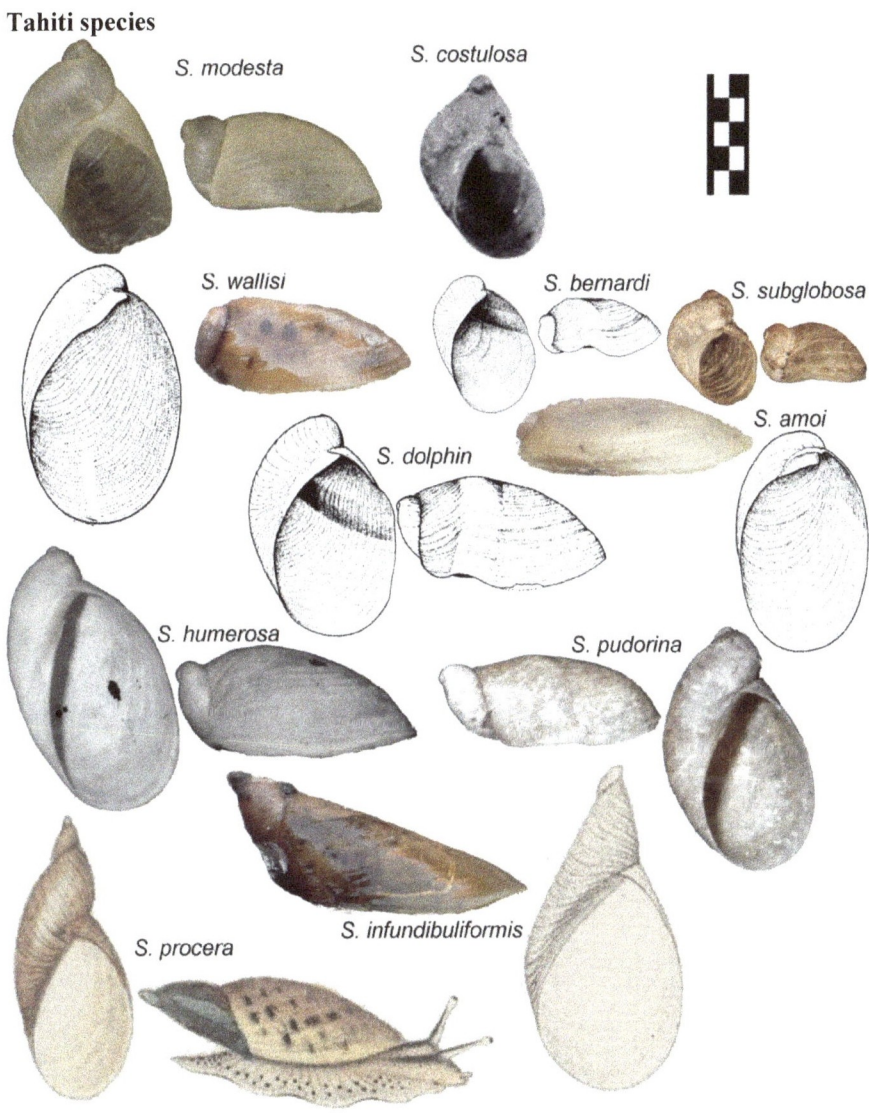

Succinea pallida Pfeiffer, 1847
   Succinea pallida Pfeiffer, 1846
   Raiatea, Tahaa. Historically very abundant in moist places on ground, now extinct. Photo NHMUK (Cuming collection).
Succinea papillata Pfeiffer, 1850
   Succinea papillata Pfeiffer, 1850
   Succinea labiata Pease, 1868
   Raiatea (moist ground in upper parts of valleys). Historically uncommon, now probably extinct. Figure from Pfeifer 1850.
Succinea procera Gould, 1846
   Succinea procera Gould, 1846
   Tahiti, Moorea. Garrett (1884) thought this an elongate *S. pudorina*. Figure from Gould 1846.
Succinea pudorina Gould, 1846
   Succinea pudorina Gould, 1846
   Succinea Gouldiana Pfeiffer, 1850
   Succinea De Gagei Garrett, 1879
   Tahiti, Moorea (Mt. Tohiea), historically the most abundant species (Garrett 1884), on trees and foliage. Photo NHMUK 78.1.28.498
Succinea subglobosa Garrett, 1884
   Succinea subglobosa Garrett, 1884
   Tahiti (formerly common on tree trunks). Photo lectotype ANSP 23646.
Succinea tahitensis Pfeiffer, 1847
   Succinea Tahitensis Pfeiffer, 1846

Moorean species

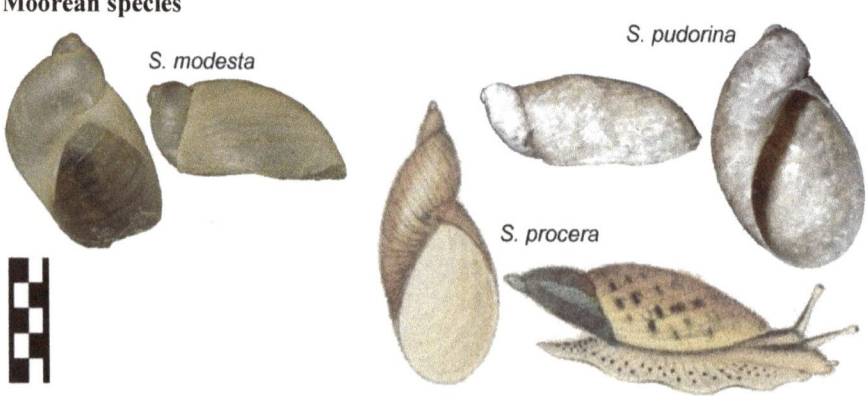

Succinea papillata Carpenter (non Pfeiffer), 1864
Huahine – historically abundant on the ground in most places, now Extinct. Last recorded in 1987 (Murray field notes). Photo NHMUK.

Succinea wallisi Cooke & Clench, 1945
Succinea wallisi Cooke & Clench, 1945
Tahiti (Papenoo, Mt. Marau, Mt. Aorai, Mt. Orofena, Taiarapu – Pueu valley). Historically from near sea level in Papenoo, now restricted to higher altitudes (above 1000m). Figure from Cooke & Clench 1945.

Raiatea, Tahaa and Huahine species

S. *tahitensis*   S. *pallida*   S. *papillata*

S. *wallisi*   S. *procera*   S. *infundibuliformis*   S. *humerosa*   S. *pudorina*

Orthurethra
Superfamily Partuloidea
Partulidae

The largest family in the Society islands is the Partulidae, with 55 species of *Partula* and *Samoana*. Due to the catastrophic introduction of *Euglandina rosea*, only 21-25 species survive today. The family was comprehensively reviewed recently (Gerlach 2016) so only brief summaries are presented here.

Partula otaheitana (Bruguière, 1792)
Tahiti – survives at higher altitudes.
Partula laevigata (Pfeiffer, 1856)
Tahiti – may survive on Teihomono plateau.
Partula jackieburchi (Kondo, 1981)
Tahiti – may survive at high altitudes.
Partula producta (Pease, 1865)
Tahiti – extinct.
Partula affinis Pease, 1868
Tahiti – may survive in south-east, being reintroduced elsewhere.
Partula compressa Reeve, 1850
Tahiti – may survive at higher altitudes.
Partula cytherea Cooke & Crampton, 1930
Tahiti – may survive at higher altitudes.

Tahiti species

Partula nodosa Pfeiffer, 1853
Tahiti – being reintroduced.
Partula diminuta C.B. Adams, 1851
Tahiti – extinct.
Partula clara Pease, 1865
Tahiti – survives in lowlands.
Partula hyalina Broderip, 1832
Tahiti – survives in lowlands.
Partula incrassa (Crampton, 1916)
Tahiti – survives in Tiapa valley, extinct on Moorea.

Moorean species

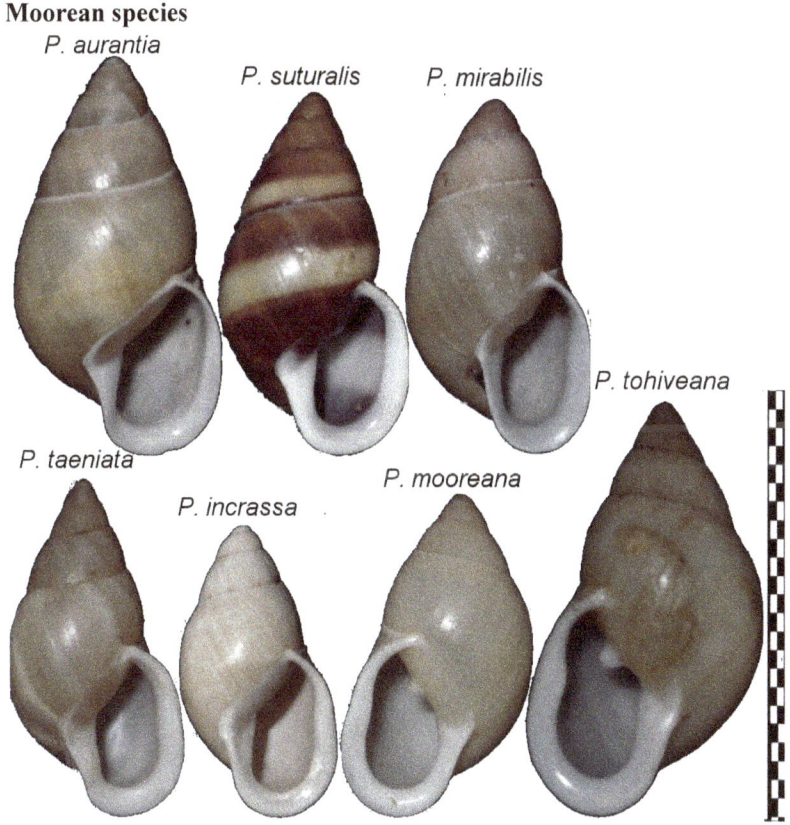

P. aurantia
P. suturalis
P. mirabilis
P. tohiveana
P. taeniata
P. incrassa
P. mooreana

Partula suturalis Pfeiffer, 1855
Moorea – being reintroduced.
Partula aurantia Crampton, 1932
Moorea – extinct.
Partula mirabilis Crampton, 1924
Moorea – being reintroduced.
Partula taeniata (Morch, 1850)
Moorea – survives.
Partula mooreana Hartman, 1880
Moorea – being reintroduced.
Partula tohiveana Crampton, 1924
Moorea – being reintroduced.
Partula varia Broderip, 1832)
Huahine – being reintroduced.
Partula rosea Broderip, 1832
Huahine – being reintroduced.
Partula arguta (Pease, 1866)
Huahine – extinct.
Partula faba (Gmelin, 1791)
Raiatea and Tahaa – extinct.
Partula navigatoria (Pfeiffer, 1849)
Raiatea – being reintroduced.
Partula dentifera Pfeiffer, 1853
Raiatea – extinct.

Huahine species

P. rosea   P. varia   P. arguta

Partula auriculata Broderip, 1832
Raiatea – extinct.
Partula protracta Crampton, 1956
Raiatea – extinct.

Raiatea species

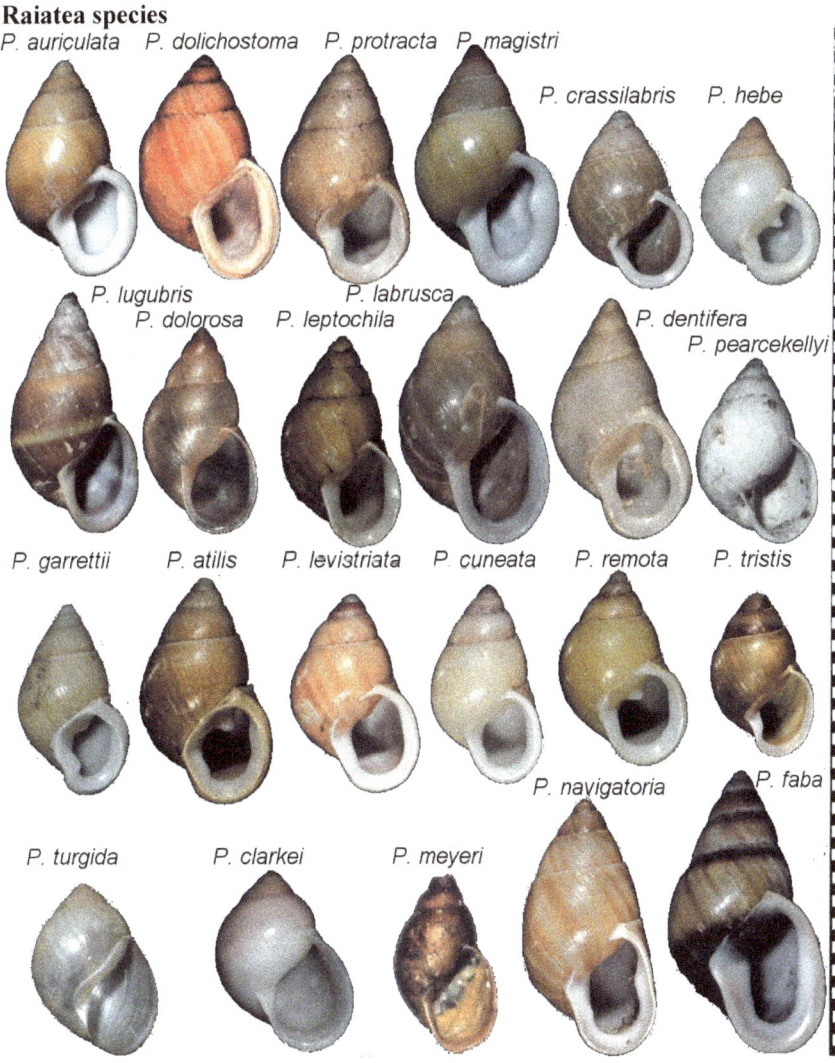

Partula dolichostoma Crampton, 1956
Raiatea – extinct.
Partula magistri Gerlach, 2016
Raiatea – extinct.
Partula lugubris Pease, 1865
Raiatea – extinct.
Partula leptochila Crampton, 1956
Raiatea – extinct.
Partula labrusca Crampton, 1953
Raiatea – extinct.
Partula dolorosa Crampton & Cooke, 1953
Raiatea – extinct.
Partula garrettii Pease, 1865
Raiatea – being reintroduced.
Partula cuneata Crampton, 1956
Raiatea – extinct.
Partula levistriata Crampton, 1956
Raiatea – extinct.
Partula remota Crampton, 1956
Raiatea – extinct.
Partula atilis Crampton, 1956
Raiatea – extinct.
Partula tristis Crampton & Cooke, 1953
Raiatea – extinct.
Partula hebe (Pfeiffer, 1846)
Raiatea – being reintroduced.
Partula crassilabris Pease, 1866
Raiatea – extinct.
Partula pearcekellyi Gerlach, 2016
Raiatea – extinct.
Partula turgida (Pease, 1865)
Raiatea – extinct.
Partula clarkei Gerlach, 2016
Raiatea – extinct.
Partula meyeri Burch, 2007
Raiatea – extinct.
Partula planilabrum Pease, 1864
Tahaa – extinct.

Partula umbilicata Pease, 1866
Tahaa – extinct.
Partula bilineata Pease, 1866
Tahaa – extinct.
Partula eremita Crampton & Cooke, 1953
Tahaa – extinct.
Partula sagitta Crampton & Cooke, 1953
Tahaa – extinct.
Partula lutea Lesson, 1831
Bora Bora and Maupiti – extinct.
Samoana attenuata (Pease, 1865)
Tahiti, Moorea, Raiatea, Tahaa, Bora Bora – survives on the first three islands.
Samoana annectens (Pease, 1865)
Huahine – survives.

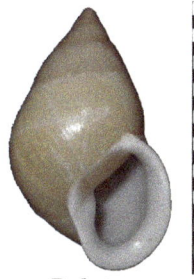

P. lutea

Tahaa species

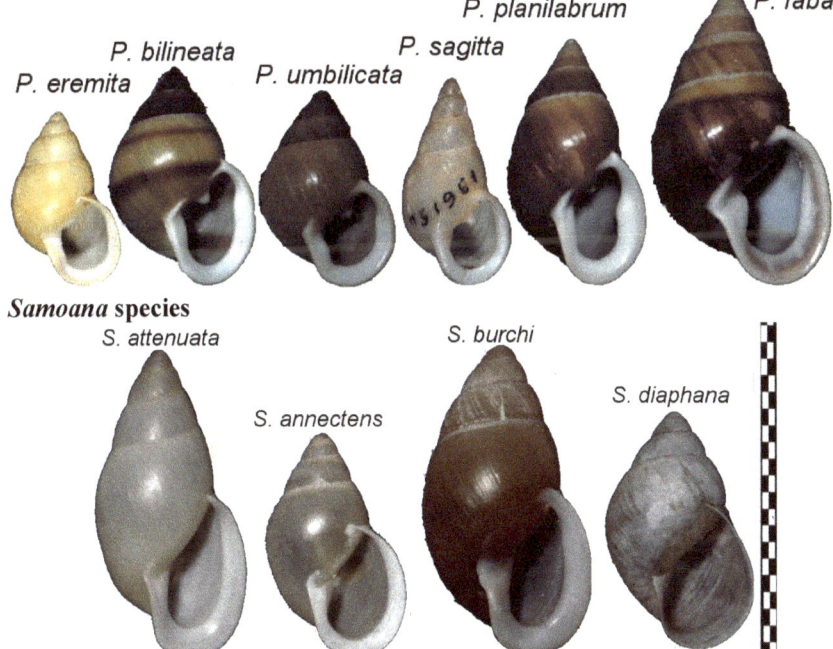

P. eremita, *P. bilineata*, *P. umbilicata*, *P. sagitta*, *P. planilabrum*, *P. faba*

Samoana species

S. attenuata, *S. annectens*, *S. burchi*, *S. diaphana*

Samoana diaphana (Crampton & Cooke, 1953)
Tahiti, Moorea – survives.
Samoana burchi Kondo, 1973
Tahiti – survives.

Superfamily Achatinelloidea
Achatinellidae
Elasmius apertum (Pease, 1865)
   Tornatellina aperta Pease, 1865
   Elasmias apertum (Pease) Cooke & Kondo, 1961
   Pacific – Tahiti, Moorea, Huahine, Raiatea, Tahaa, Bora Bora, Maupiti. Locally common on vegetation.
Elasmius peaseanum (Garrett, 1884)
   Tornatellina peaseana Garrett, 1884
   Elasmias peasianum (Garrett) Cooke & Kondo 1961
   Tahiti, Moorea. Very rare on foliage. Photos lectotype ANSP 24023.
Lamellidea (Lamellidea) oblonga (Pease, 1865)
   Lamellidea (Lamellidea) oblonga (Pease) Cooke & Kondo, 1961
   Pacific - Mehetia, Tahiti, Moorea, Huahine, Raiatea, Tahaa, Bora Bora, Maupiti. Locally abundant. Figure from Cooke & Kondo 1961.
Lamellidea (Lamellidea) pusilla (Gould, 1847)
   Tornatellina serrata Pease Garrett 1884
   Lamellidea (Lamellidea) pusilla (Gould) Cooke & Kondo 1961
   Pacific - Mehetia, Tahiti, Moorea, Huahine, Raiatea, Tahaa, Bora Bora, Maupiti. '*Tornatellina impressa intuscostata*' in BPBM are immature of this species (Cooke & Kondo, 1961), as are '*Tornatellina conica*' labeled by Garrett.
Lamellidea (Lamellidea) micropleura Cooke & Kondo, 1961
   Lamellidea (Lamellidea) micropleura Cooke & Kondo, 1961
   Moorea (Faatoai, Teaharoa under bark and lichen). A similar species from Pitcairn (Preece 1995). Figure from Cooke & Kondo 1961.
Tornatellides (Tornatellides) oblongus (Anton, 1839)
   Tornatellina oblonga Pease 1864
   Tornatellidea simplex Pease, 1865

Elasmias apertum

Tornatellidea oblongus (Anton) Cooke & Kondo, 1961
Pacific (Cooks to Pitcairn) - Mehetia, Tahiti, Moorea, Huahine, Raiatea, Tahaa, Bora Bora, Maupiti. Common up to 150m, rarely to 300m (Garrett 1884). Figure from Cooke & Kondo 1961.

Tornatellinops philippii (Pfeiffer, 1849)
Tornatellina philippii Pfeiffer, 1849
Tornatellinops philippii (Pfeiffer) Cooke & Kondo 1961
Cooks, Australs, Marquesas – Tahiti (under stones, leaves, wood). Photo NHMUK.

Tubuaia raoulensis Pilsbry & Cooke, 1915
Tornatellina societatis Pilsbry & Cooke, 1915
Australs – Raiatea (Motu Iriru). Photo *T. societatis* lectotype ANSP 83155.

Tubuaia voyana Pilsbry & Cooke, 1915 **peasei** Kondo, 1962
Tornatellina perplexa Garrett, 1879 misidentified
Tornatellina nitida Pease Garrett 1884
Tubuaia voyana peasei Kondo, 1962

Huahine and 'Not uncommon, and ranges throughout the group' (Garrett 1884). The species is also found in the Cooks and Austral islands. Figure from Cooke & Kondo 1961.

Superfamily Pupilloidea
Valloniidae
Pupisoma orcula (Benson, 1850)
Moorea – (Opunohu - Gargominy 2010), Raiatea (1992).

Vertiginidae (including 'Gastrocoptidae')
Probably misidentified or mislabeled specimens have been called *Pronesopupa* ? sp. (Tahiti – Fautaua 1926-34, and Mehetia 1934) and *Pupina complanata* Pease ('Tahiti'). The former is the only edentulous vertiginid genus and may be a mistake for young of other species.

Gastrocopta (Sinalbinula) pediculus (Shuttleworth, 1852)
Vertigo hyalina Zel. Pease 1871
Pupa hyalina "Zelebor" Pfeiffer 1868
Pupa pediculus Shuttleworth Boettger 1880
Vertigo pediculus (Shuttleworth) Garrett 1884
Gastrocopta pediculus (Shuttleworth) Pilsbry 1916
Pacific (introduced from south-east Asia?) - Mehetia, Tahiti, Moorea, Huahine, Raiatea (Motu Iriru 1992), Tahaa, Maupiti. Present since the 19th century (Cuming specimens from Tahiti). Figure from Pilsbry 1916.

Gastrocopta servilis (Gould, 1843)
Introduced from Carribean? - Mehetia (1934, as *lyonisiana* in BPBM). Figure from Pilsbry 1916.

Nesopupa (Nesopupa) armata (Pease, 1871)
Vertigo armata Pease, 1871
Pupa tantilla var. *armata* Pease Boettger 1880
Nesopupa armata (Pease) Pilsbry 1918-20
Cooks, Makatea - Mehetia, Huahine, Bora Bora. Photo lectotype MCZ 48315.

Nesopupa (Nesopupa) pleurophora (Shuttleworth, 1852)
Pupa pleurophora Shuttleworth, 1852
Pupa Dunkeri "Zelebor" Pfeiffer, 1868
Vertigo dunkeri Zel. Pease 1871
Pupa tantilla var. *pleurophora* Shuttleworth Boettger 1880
Nesopupa pleurophora (Shuttleworth) Pilsbry 1918-20
Makatea - Tahiti (Mt. Aorai 1000m), Moorea. Figure from Pilsbry 1916.

Nesopupa (Nesopupa) tantilla (Gould, 1847)
    Pupa (Vertigo) tantilla Gould, 1847
    Vertigo tantilla (Gould) Gould 1852
    Nesopupa tantilla (Gould) Pilsbry 1918-20
    Pacific (non-Societies specimens of uncertain identification) - Tahiti, Huahine, Raiatea, Bora Bora, Maupiti. Photo holotype USNM 5505.

Nesoropupa **species**
    Endemic genus of elongate species (except *N. duodecim*) on Tahiti: on both Mts. Aorai and Marau (1430-2065m): ***N. duodecim*** Gargominy, 2008, ***N. fontainei*** Gargominy, 2008 and ***N. nathaliae*** Gargominy, 2008; or Mt. Aorai (2065m) only ***N. fenua*** Gargominy, 2008.

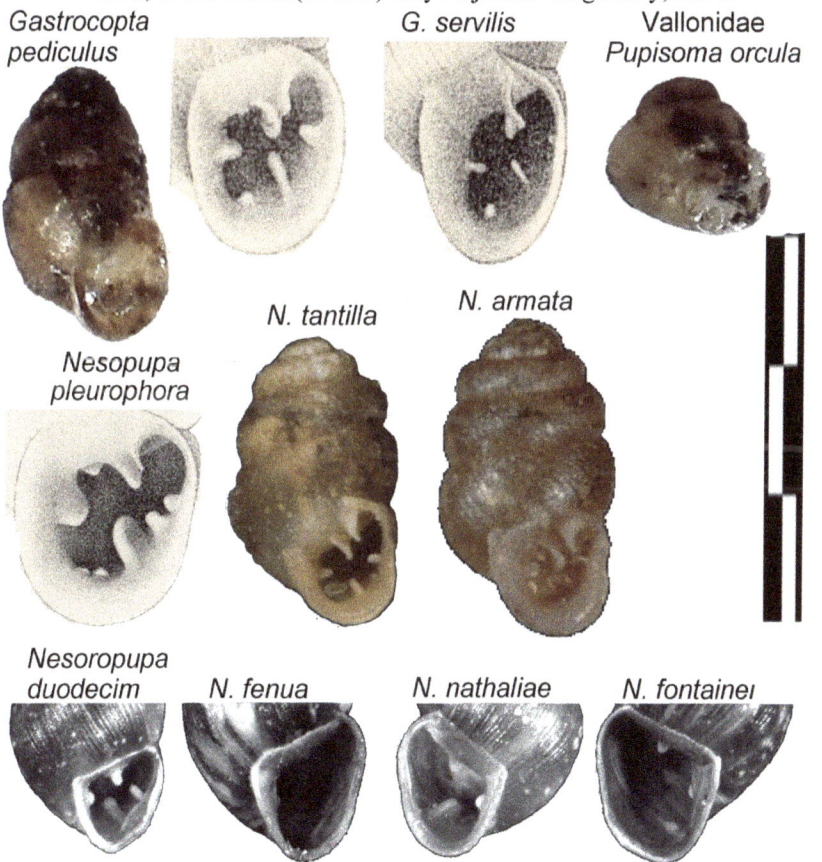

Sigmurethra
Superfamily Helicoidea
Helicidae
Cornu aspersa (Muller, 1774)
   Helix (Cryptomphalus) aspersa (Muller) Solem 1964
   Introduced – cited from Tahiti by Solem (1964), possibly in error.

Bradybaenidae
Bradybaena similaris (Rang, 1831)
   Bradybaena similaris (Rang) Brook 2010
   Introduced – Tahiti, Moorea, Huahine, Raiatea, Maupiti. First recorded in 1934 (Moorea).

Superfamily Achatinoidea
Achatinidae
Lissachatina fuliuca (Ferussac, 1821)
   Achatina fulica Bowdich Tillier & Clarke 1983
   Lissachatina fulica (Ferussac) Fontanilla *et al.* 2014
   Introduced – Tahiti, Moorea, Huahine, Raiatea, Tahaa, Bora Bora, Maupiti. Deliberately introduced to Tahiti in 1967, spreading to Moorea by 1978 and other islands by 1982.

Lissachatina fulica

Bradybaena similaris

Subulinidae
   Allopeas gracile (Hutton, 1834)
      Bulimus junceus Gould, 1846
      Stenogyra tuckeri Pfeiffer Garrett 1884
      Opeas gracile Pilsbry 1906-7
      Introduced? – Mehetia, Tahiti, Moorea, Huahine, Raiatea, Tahaa, Bora Bora, Maupiti (including Motu Tiapaa). Sea level to 1,000m. First recorded in 1834 (Mehetia). Photo *B. junceus* holotype USNM 5489.
   Allopeas kyotense (Pilsbry & Hirase, 1904)
      Allopeas clavulinum (Poitiez & Michaud) Brook 2010
      Introduced – Tahiti, Moorea, Raiatea, Bora Bora. First recorded in 1925 (Brook 2010).
   Lamellaxis micra (d'Orbingy, 1835)
      Introduced - Moorea (Belvedere only). First recorded in 2010 (specimens in FLMNH).
   Leptinaria unilamellata (d'Orbingy, 1837)
      Leptinaria unilamellata (d'Orbingy) Kahn *et al.* 2014
      Introduced – Tahiti, Moorea, Huahine, Raiatea, Tahaa. First recorded in 1991 but introduced earlier as shells have been found in archaeological deposits (Kahn *et al.* 2014).

Paropeas achatinaceum (Pfeiffer, 1846)
Introduced – Tahiti, Moorea, Huahine, Raiatea, Tahaa. First recorded in 1927 (Tahiti).

Opeas hannense (Rang, 1831)
Opeas hannense (Rang) Brook 2010
Introduced – Tahiti, Moorea, Raiatea (including Motu Iriru). First recorded in 1925 (Tahiti, Moorea).

Subulina octona Bruguierie, 1789
Subulina octona Bruguierie Boettger 1909
Introduced – Tahiti, Moorea, Huahine, Raiatea, Bora Bora, Maupiti. First recorded in 1909 (Tahiti), present on all other islands by 1925.

Superfamily Testacelloidea
Spiraxidae

Euglandina 'rosea' Ferussac, 1821
Euglandina rosea (Ferussac) Tillier & Clarke 1983
Introduced – Tahiti, Moorea, Huahine, Raiatea, Tahaa, Bora Bora, Maupiti. Deliberately introduced in 1974 (Tahiti), with subsequent releases on Moorea (1977). Introductions to other islands are undated (estimated around 1985). Last introduction to Huahine around 1993. Comprises more than one species (Meyer *et al.* 2017).

Superfamily Streptaxoidea
Streptaxidae

Streptostele (Tomostele) musaecola (Morelet, 1860)
Streptostele (Tomostele) musaecola (Morelet) Brook 2010
Introduced – Tahiti, Moorea, Huahine, Raiatea (including Motu Iriru), Maupiti. First recorded in 1977 (Brook 2010).

Subulina octona   *Paropeas achatinaceum*   *Allopeas gracile*

Superfamily Punctoidea
Charopidae
- ***Sinployea lamellicosta*** (Garrett, 1884)
  - *Patula lamellicosta* Garrett, 1884
  - *Helix (Patula) lamellicosta* (Garrett) Tryon 1887
  - *Endodonta lamellicosta* (Garrett) Pilsbry 1893
  - *Sinployea lamellicosta* (Garrett) Solem 1983
  - Tahiti (Pirae-Aorai trail and to west at 1,600m), a scarce species found under rotten wood (Garrett 1884). Photo NMHUK.
- ***Sinployea modicella*** (Ferussac in Deshayes, 1840)
  - *Helix modicella* Ferussac, 1840
  - *Pitys modicella* (Ferussac) Pease 1871
  - *Patula modicella* (Ferussac in Deshayes) Garrett 1884
  - *Endodonta (Charopa) modiella* (Ferussac) Pilsbry 1893
  - *Sinployea modicella* (Ferussac in Deshayes) Solem 1983
  - Moorea (Opunohu, Faatoai, Maramu, Tepatu, Belvedere, Mt Tohiea, in litter and on trees, under dead wood in lowlands). Photo NMHUK.
- ***Sinployea montana*** Solem, 1983
  - *Sinployea montana* Solem, 1983
  - Tahiti (west of Mt. Aorai trail at 2,000m). Figure from Solem (1983).
- ***Sinployea neglecta*** Solem, 1983
  - *Sinployea neglecta* Solem, 1983
  - Huahine (near Tiva and Fare, on logs and dead leaves at 350-400m). Figure from Solem (1983).

Euglandina rosea   *Streptostele muscaeola*

Sinployea tahitiensis Solem, 1983
> *Sinployea tahitiensis* Solem, 1983
> Tahiti (Fautaua, Papenoo, Mt. Aorai, Mt. Marau, historically above, now still fairly common above 1,050m). Found in *Freycinetia* axils, probably other habitats as well. Figure from Solem (1983).

***Sinployea* sp.**
> *Sinployea* sp. Solem 1983Bora Bora (south slope of Pahio-Temanu ridge, 250m). Known from one subadult specimen (BPBM 152395).

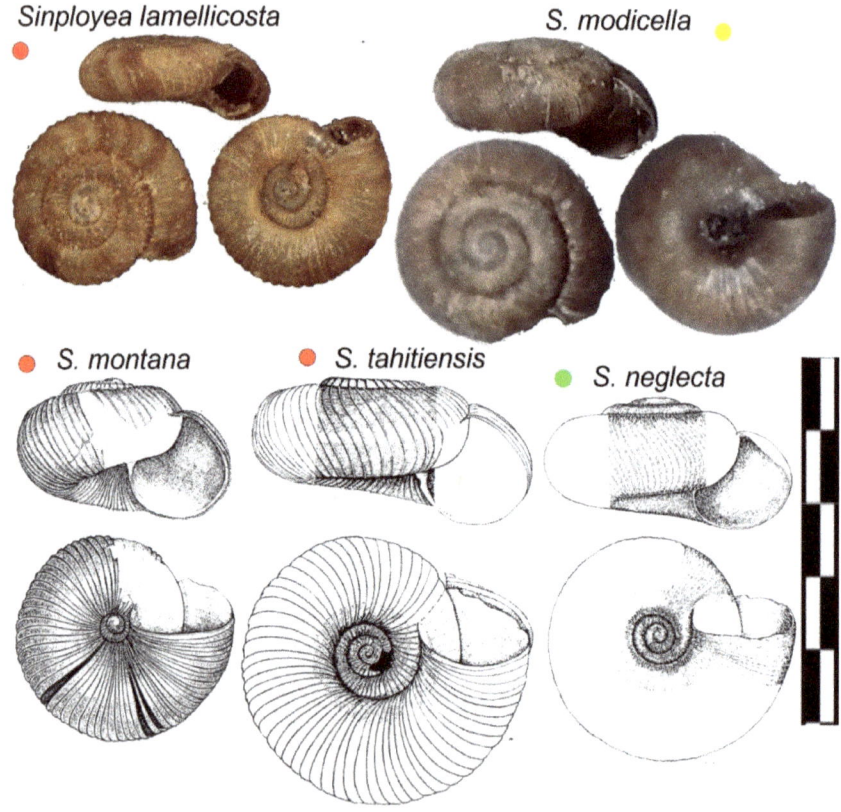

Endodontidae
    Libera bursatella (Gould, 1846)
        Helix bursatella Gould, 1846
        Helix turricula Hombron & Jacquinot, 1848
        Libera bursatella (Gould) Garrett 1884
        Libera bursatella oroferensis Solem, 1982
        Tahiti (Mts Aorai, Orofena above 1,300m, in *Freycinetia* axil). *Orofenensis* from Mt. Orofena. Figure from Solem (1982).
    Libera cookeana Solem, 1982
        Libera cookeana Solem, 1982
        Tahiti (Mt. Aorai above 1,800m). Figure from Solem (1982).
    Libera dubiosa (Ancey, 1889)
        Libera dubiosa Ancey, 1889
        Libera coarctata Garrett, 1884 (not Pfeiffer)
        Moorea (on the ground in north and east, not collected since 1800s – Kahn *et al.* 2014). Figure from Solem (1982).
    Libera garrettiana Solem, 1982
        Libera heynemanni Garrett 1884 (non Pfeiffer)
        Helix (Libera) heynemanni Tryon 1887 (non Pfeiffer)
        Libera garrettiana Solem, 1982
        Tahiti (very abundant in north-west, on ground - Garrett 1884, now Mt. Marau 1320m only. Figure from Solem 1982.
    Libera gregaria Garrett, 1884
        Libera gregaria Garrett, 1884
        Moorea (two valleys in south-west, in 'immense numbers' under stones – Garrett 1884). Photo lectotype ANSP 47825.
    Libera heynemanni (Pfeiffer, 1862)
        Helix bursatella Gould, 1846 (part)
        Helix Heynemanni Pfeiffer, 1862
        Pitys Heynemanni Pease 1871
        Patula Heynemani Schmeltz 1874
        Libera heynemanni Pfeiffer Garrett 1884
        Probably Tahiti. Figure from Solem (1982).
    Libera incognata Solem, 1982
        Libera incognata Solem 1982
        Probably Tahiti. Figure from Solem (1982).
    Libera jacquinoti(Pfeiffer, 1850)
        Helix jacquinoti Pfeiffer, 1850
        Libera jacquinoti (Pfeiffer) Solem 1982

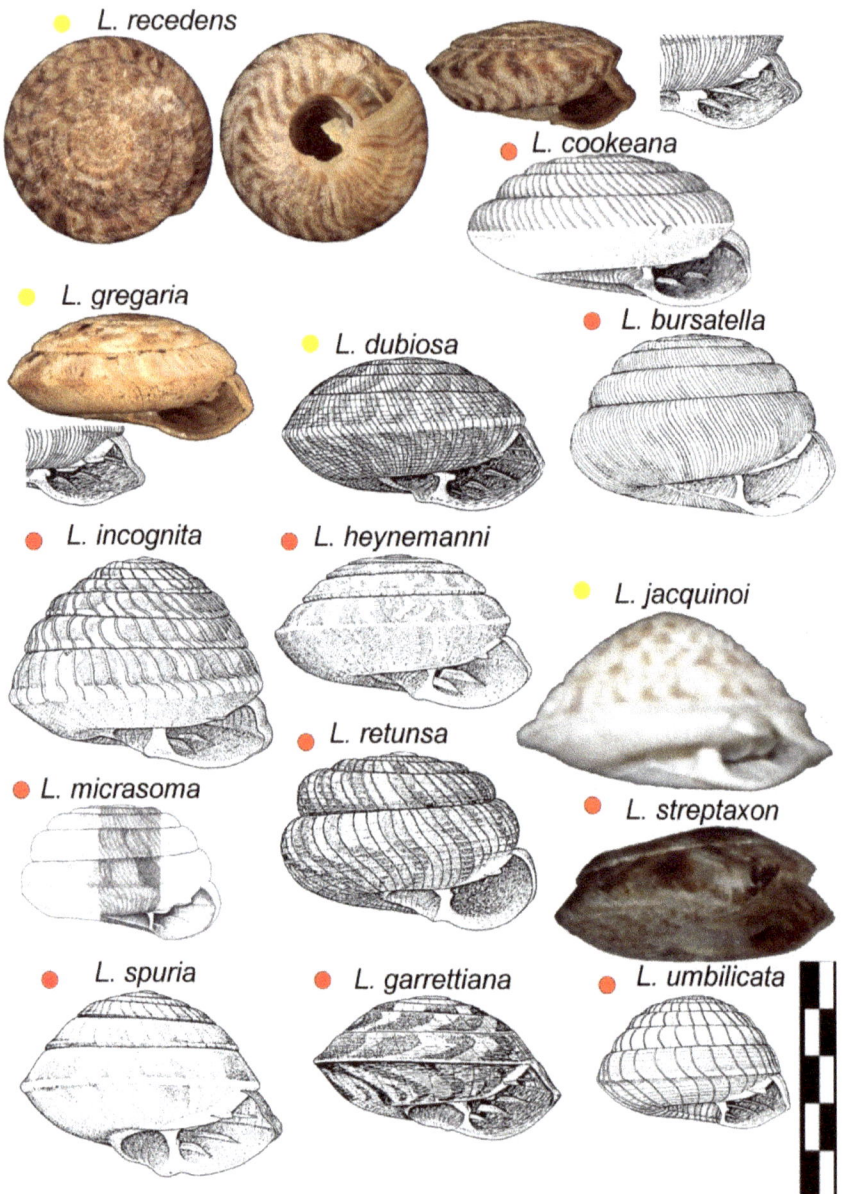

Moorea (not recorded since 1839 except in archaeological deposits – Kahn *et al.* 2014). Photo lectotype NHMUK 1962707/1

Libera micrasoma Solem, 1982
Libera micrasoma Solem, 1982
Tahiti (Mt. Aorai above 1,800m. Figure from Solem (1982).

Libera recedens Garrett, 1884
Libera recedens Garrett, 1884
Moorea (lower parts of one valley in west, 'very abundant beneath decaying vegetation' Garrett 1884). Photo lectotype ANSP 47827.

Libera retunsa (Pease, 1864)
Helix retunsa Pease, 1864
Pitys retunsa Pease 1871
Libera retunsa Pease Garrett 1884
Tahiti (south, not uncommon under rotten wood – Garrett 1884). Figure from Solem (1982).

Libera spuria (Ancey, 1889)
Libera heynemanni var. spuria Ancey, 1889
Libera spuria (Ancey) Solem 1982
Origin uncertain - Tahiti? Figure from Solem (1982).

Libera streptaxon (Reeve, 1852)
Helix bursatella Gould, 1846
Helix coarctata Pfeiffer, 1849
Helix streptaxon Reeve, 1852
Libera coarctata Pfeiffer Garrett 1884
Helix turricula Hambron & Jacquinot, 1852
Libera streptaxon Reeve Solem, 1982
Origin uncertain - Tahiti?. Photo syntype NHMUK 1962708.

Libera umbilicata Solem, 1982
Libera umbilicata Solem, 1982
Tahiti (Orofena). Figure from Solem (1982).

Mautodontha (Mautodontha) aoraiensis Solem, 1982
Mautodontha (M.) aoraiensis Solem, 1982
Tahiti (Mt. Aorai, Mt. Marau, above 1,300m, rare on trees). Figure from Solem (1982).

Mautodontha (M.) boraborensis (Garrett, 1884)
Pitys boraborensis Garrett, 1884
Mautodontha (Mautodontha) boraborensis (Garrett) Solem 1982
Bora Bora at 250m. Photo lectotype ANSP 47775.

Mautodontha (M.) zimmermani Solem, 1982
    Mautodontha (M.) zimmermani Solem, 1982
    Tahiti (Mt. Aorai). Figure from Solem (1982).
Mautodontha (Garrettoconcha) acuticosta (Garrett, 1884)
    Patula acuticosta Garrett, 1884
    Mautodontha (Garrettoconcha) acuticosta (Garrett) Solem 1982
Raiatea, rarer than *M. consimilis*. Figure from Solem (1982).
Mautodontha (Garrettoconcha) consimilis (Pease, 1868)
    Helix consimilis Pease, 1868
    Pitys consimilis Pease 1871
    Patula consimilis Pease Garrett 1884
    Mautudontha (Garrettoconcha) consimilis (Pease) Solem 1982
    Raiatea in all larger valleys (Garrett 1884), also listed from Tahiti but probably in error. Photo paralectotype BMNH 1962710/2
Mautodontha (Garrettoconcha) consobrina (Garrett, 1884)
    Pitys consobrina Garrett, 1884
    Mautodontha (Garrettoconcha) consobrina (Garrett) Solem 1982
    Huahine, rare and restricted to one valley (not specified) (Garrett 1884). Figure from Solem (1982).
Mautodontha (Garrettoconcha) maupiensis (Garrett, 1872)
    Pitys maupiensis Garrett, 1872
    Mautodontha (Garretoconcha) maupiensis (Garrett) Solem 1982
    Maupiti, very common (Garrett 1884). Figure from Solem (1982).
Mautodontha (Garrettoconcha) parvidens (Pease, 1861)
    Helix parvidens Pease, 1861
    Pitys parvidens Pease 1871
    Mautodontha (Garrettoconcha) parvidens (Pease) Solem 1982
    Tahiti, Moorea, Huahine, very abundant (Garrett 1884). Figure from Solem (1982).
Mautodontha (Garrettoconcha) punctiperforata (Garrett, 1884)
    Pitys punctiperforata Garrett, 1884
    Mautodontha (Garrettoconcha) punctiperforata (Garrett) Solem 1982
    Moorea. Figure from Solem (1982).
Mautodontha (Garrettoconcha) saintjohni Solem, 1982
    Mautodontha (Garrettoconcha) saintjohni Solem, 1982
    Bora Bora at 200m. Figure from Solem (1982).
Mautodontha (Garrettoconcha) subtilis (Garrett, 1884)
    Pitys subtilis Garrett, 1884
    Mautodontha (Garrettoconcha) subtilis (Garrett) Solem 1982

Huahine, rare and confined to "a valley on the north end" (Garrett 1884). Figure from Solem (1982).

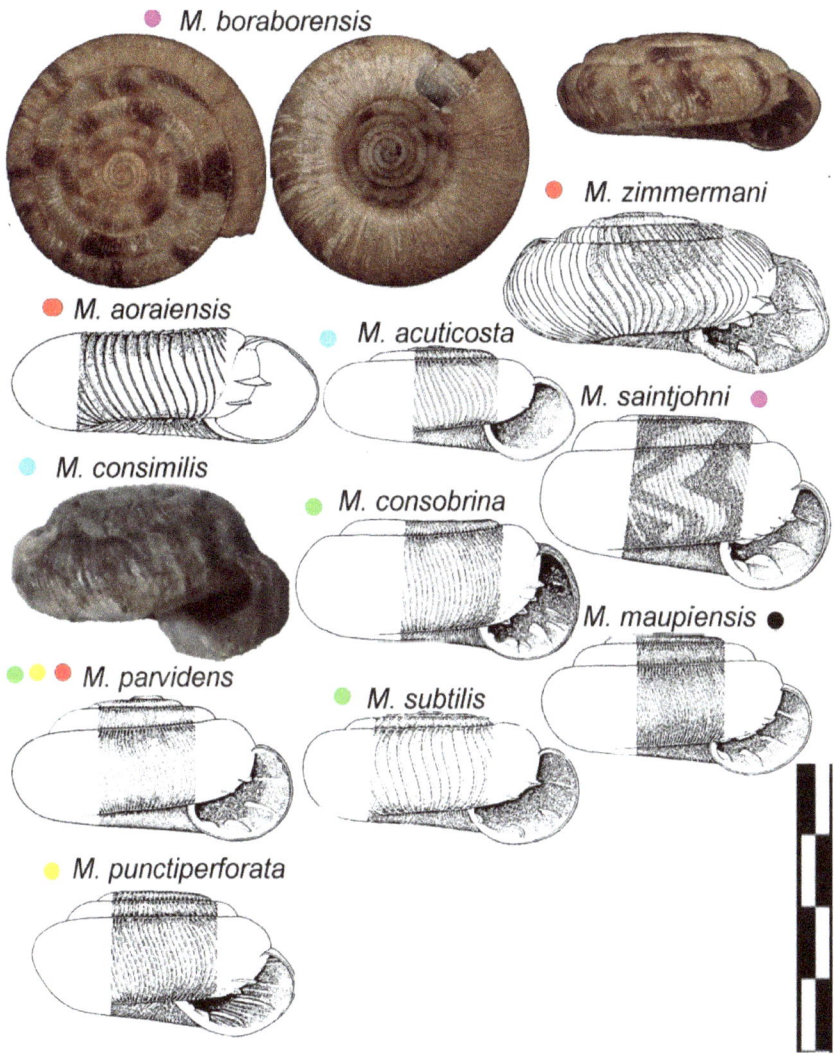

Nesodiscus cretaceus (Garrett, 1884)
  Pitys ficta Garrett (non Pease) Schmeltz 1874
  Endodonta cretacea Garrett, 1884
  Nesodiscus cretaceus (Garrett) Solem 1982
  Bora Bora, common on the ground but local at 200m (Garrett 1884). Photo lectotype ANSP 47832.

Nesodiscus fictus (Pease, 1864)
  Helix ficta Pease 1864
  Endodonta ficta Pease 1871
  Nesodiscus fictus (Pease) Solem 1982
  Tahaa (plentiful according to Garrett, but now probably extinct). Figure from Solem (1982).

Nesodiscus fabrefactus (Pease, 1864)
  Helix fabrefacta Pease, 1864
  Endodonta fabrefacta Pease 1871
  Patula fabrefacta Schmeltz 1874
  Helix conicava Pfeiffer, 1876 (nomen nudum)
  Endodonta fabrefacta var. *picea* Garrett, 1884
  Nesodiscus fabrefacta (Pease) Solem 1982
  Raiatea (four large valleys, var. *picea* Garrett, 1884 from west side), Tahaa (east coast, Pu Rauti). Figure from Solem (1982).

Nesodiscus huaheinensis (Pfeiffer, 1853)
  Endodonta huaheinensis Pfeiffer, 1853
  Helix Huaheinensis Pfeiffer, 1852
  Patula Huaheinesis Schmeltz 1874
  Nesodiscus huaheinensis (Pfeiffer) Solem 1982
  Huahine (in all large valleys – Garret 1884), one specimen collected in the 20th century, from Tiva (Solem 1982). Figure from Solem (1982).

Nesodiscus magnificus Solem, 1982
  Nesodiscus magnificus Solem, 1982
  Bora Bora at 250m. Figure from Solem (1982).

Nesodiscus obolus (Gould, 1846)
  Helix (Pitys) obolus Gould, 1846
  Endodonta obolus (Gould) Martens 1860
  Pitys obolus (Gould) Adams & Adams 1858
  Helix acetabulum Pease, 1861
  Pithys? celsa Pease, 1870
  Endodonta acetabulum Pease 1871
  Patula obolus Schmeltz 1874

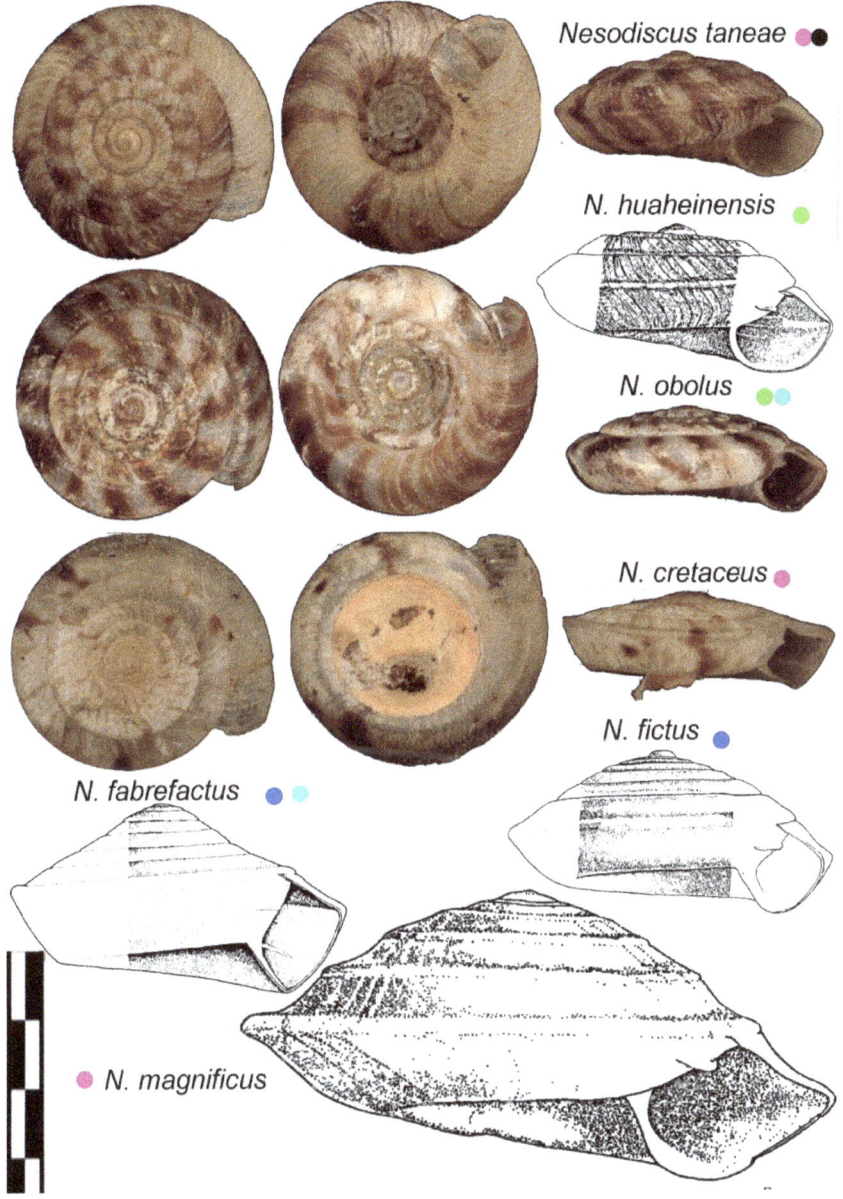

Helix celsa (Pease) Pfeiffer 1876
Patula Barffi 'Garrett' Schmeltz 1874
Patula intermedia 'Mousson' Schmeltz 1874
Nesodiscus obolus (Gould) Soelm 1976
Huahine, Raiatea, historically widespread. Photo *H. acetabulum* lectotype ANSP 47844.

Nesodiscus taneae (Garrett, 1872)
Pitys Tanneae Garrett, 1872
Patula Janeae Schmeltz, 1874
Helix Janeae (Garrett) Pfeiffer 1876
Endodonta taneae Garrett Garrett 1884
Helix Boraborensis 'Pease' Garrett 1884
Helix (Endodonta) taneae (Garrett) Tryon 1887
Enddonta garrettii Ancey, 1889
Nesodiscus taneae (Garrett) Solem 1976
Bora Bora, Maupti. Garrett (1884) recorded it to be very abundant on the ground at 200-300m.. Photo lectotype ANSP 47846.

Superfamily Gastrodontoidea
Euconulidae
Coneuplecta (Durgellina) calculosa (Gould, 1852)
Helix calculosa Gould 1852
Trochonanina calculosa (Gould) Garrett 1884
Coneuplecta calculosa (Gould) Baker 1941
Pacific - Tahiti, Moorea, Huahine. Garrett (1884) recorded it from foliage from the shore to 300m. Photo syntype USNM 5465.

Euconulus (Monoconulus) conoides Baker, 1941
Euconulus (Monoconulus) conoides Baker, 1941
Tahiti (Mt. Aorai above 1,800m).

Coneuplecta calculosa ●●●   *Euconulus conoides* ●

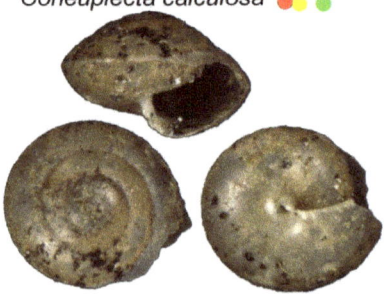

Diastole (Diastole) conula (Pease, 1861)
 Helix conula Pease, 1861
 Trochonania conula (Pease) Garrett 1884
 Diastole (Diastole) conula (Pease) Baker 1938
 Pacific – Mehetia, Tahiti, Moorea, Huahine, Raiatea, Bora Bora. Widespread and locally abundant. There are light and dark forms, always darker than *D. necrodes*. Photo paralectotype ANSP 49339.

Diastole (Diastole) necrodes Baker, 1938
 Diastole (Diastole) necrodes Baker, 1938
 Tahiti (Fautau at 200m on vegetation). This is similar to *D. conula* but is fractionally larger (diameter 7mm, compared to 6.5mm), more depressed, with shallower foveae, more carinate and a differently shaped columella.

Lamprocystis (Raiatea) simillima (Pease, 1964)
 Helix simillima Pease, 1864
 Microcystis simillima (Pease) Garrett 1884
 Lamprocystis (Raiatea) simillima (Pease) Baker 1938
 Raiatea (under leaves and wood in mountains, or on leaves above 1,000m). Specimens labeled *M. venosa* (Pease, 1866) and *L. vitrinella* (Beck, 1837) are probably misidentifications of this species. Figure from Baker (1938).

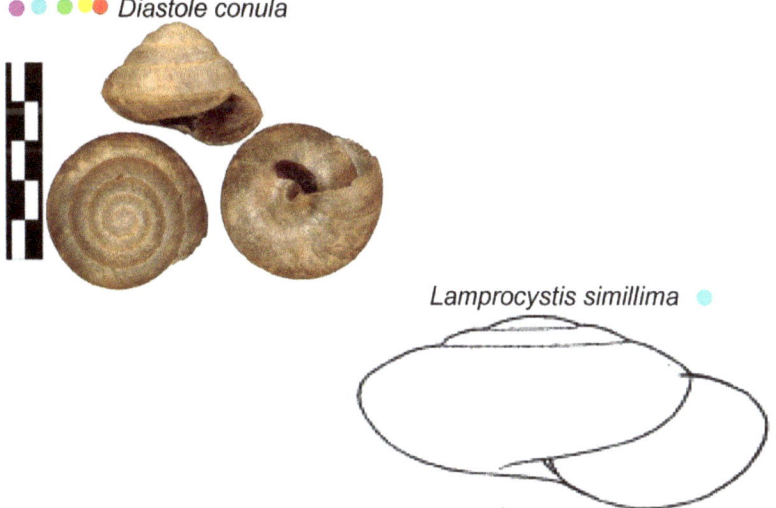

Diastole conula

Lamprocystis simillima

Hiona (Insulorbis) aorai Baker, 1940
   Hiona (Insulorbis) aorai Baker, 1940
   Tahiti (Mt. Aorai above 1,800m, in moss on trees). Figure from Baker (1940).
Hiona (Insulorbis) huahinei Baker, 1940
   Microcystis verticillata 'Pease' Garrett 1884
   Nanina verticillata Tryon 1886
   Hiona (Insulorbis) huahinei Baker, 1940
   Huahine. Figure from Baker (1940). There are unidentified *Hiona* specimens from Raiatea, Bora Bora and Maupiti in BPBM.
Hiona (Insulorbis) radians (Pfeiffer, 1853)
   Helix radians Pfeiffer, 1853
   Nanina radians Tryon 1886
   Hiona (Insulorbis) radians Pfeiffer Baker 1940
   Hiona (Insulorbis) radians pohaitare Baker, 1940
   Tahiti (Mt. Orohena, Papenoo valley, Pohaitara and Mt. Marau, above 200m, on vegetation), now only above 1,050m. *H. radians pohaitare* from Pohaitara valley and Mt. Marau is larger. Photo NHMUK.
Hiona (Minororbis) angustivoluta (Garrett, 1884)
   Microcystis angustivoluta Garrett, 1884
   Nanina angustivoluta (Garrett) Tyron 1886
   Hiona (Minorobis?) angustivoluta (Garrett) Baker 1940
   Moorea (north-east, under rotten wood). Photo lectotype ANSP 49297.
Hiona (Minororbis) cultrata (Gould, 1846)
   Helix culturata Gould, 1846
   Microcystis culturata (Gould) Garrett 1884
   Nanina culturata (Gould) Bland & Binney 1872
   Hiona (Minororbis) culturata (Gould) Baker 1940
   Tahiti (west of Aorai trail 6,000ft, 1934), possibly includes specimens from Papenoo (1927), under leaves, on vegetation. Photo NHMUK. Ancey's (1889). *Microcystis Marici* (***H. marici*** (Ancey, 1889)) from 'Tahiti' is known only from his record. Apparently similar to *H. culturata*, differing in number of whorls and size, however his measurements lie within the range of *H. culturata*, the only remaining difference is the reported presence of a spiral band.
Hiona (Minororbis) scalpta (Garrett, 1884)
   Microcystis scalpta Garrett, 1884
   Nanina tahaensis Tryon, 1886
   Hiona (Minorbis?) scalpta (Garrett) Baker 1940

Tahaa (plentiful in small area of Haamene, under stones, leaves rotten wood, not found anywhere else according to Garrett, now probably extinct). Photo *N. tahaensis* lectotype ANSP 49289.

Hiona (Minororbis) verticillata (Pease, 1868)
Helix brunnea Carpenter, 1864
Nanina verticillata Pease, 1868
Helicopsis verticillata Pease 1871
Microcystis verticillata (Pease) Garrett 1884
Hiona (Minororbis) verticillata (Pease) Baker 1940
Moorea (under leave and rotten wood in northern valleys). Specimens from supposedly from 'Tahiti' are probably mislabeled and those from Huahine are misidentifications of *H. huahinei*. Photo lectotype ANSP 49284.

Liardetia (Belonesia) undulata Baker, 1938
Liardetia (Belonesia) undulata Baker, 1938
Raiatea (Tehehani Rahi, very rare on dead *Pandanus* leaves at 600m). Figure from Baker (1938).

Liardetia (Dasyconus) decussata Baker, 1938
Liardetia (Dasyconus) decussata Baker, 1938
Liardetia (Dasyconus) decussata asperior Baker, 1938
Tahiti (from 250-2,100m, Vaipoe, Fautauam, Vaihiria and Mt. Aorai), on vegetation. *L. decussata asperior* from Mt. Aorai with a more pronounced keel). Figure from Baker (1938).

Liardetia (Dasyconus) normalis (Pease, 1865)
Helix normalis Pease, 1865
Helicopsis normalis (Pease) Pease 1871
Microcystis normalis (Pease) Garrett 1884
Nanina normalis (Pease) Tryon 1886
Liardetia normalis (Pease) Baker 1938
Liardetia normalis aequior Baker, 1938
Tahiti, Moorea (Maramu, Tohieepatua 2010), Huahine, Bora Bora. Historically very abundant on vegetation, now localised. *L. normalis aequior* from Moore (Tepatu, east Faatoai, Mt Tohiea). Figure from Baker (1938).

Liardetia (Dasyconus) perplexa H.B. Baker, 1938
Liardetia (Dasyconus) perplexa Baker, 1938
Huahine, Raiatea. On dead leaves from sea level to 600m. Figure from Baker (1938).

Liardetia (Dasyconus) subrugosa (Garrett, 1884)
> *Trochonanina subrugosa* Garrett, 1884
> *Liardetia (Dasyconus) subrugosa* (Garrett) Baker 1938
> Tahiti north-west), Moorea (Mt. Tohiea). Historically from around 300m, now only in the highest areas, under stones. Photos lectotype ANSP 49302.

Liardetia (Dasyconus) tahitensis (Garrett, 1884)
> *Trochonanina tahitensis* Garrett, 1884
> *Liardetia (Dasycinus) tahitensis* (Garrett) Baker 1938
> Tahiti (Mt. Aorai and Orofena, above 700m, on vegetation). Photo lectotype ANSP 49303.

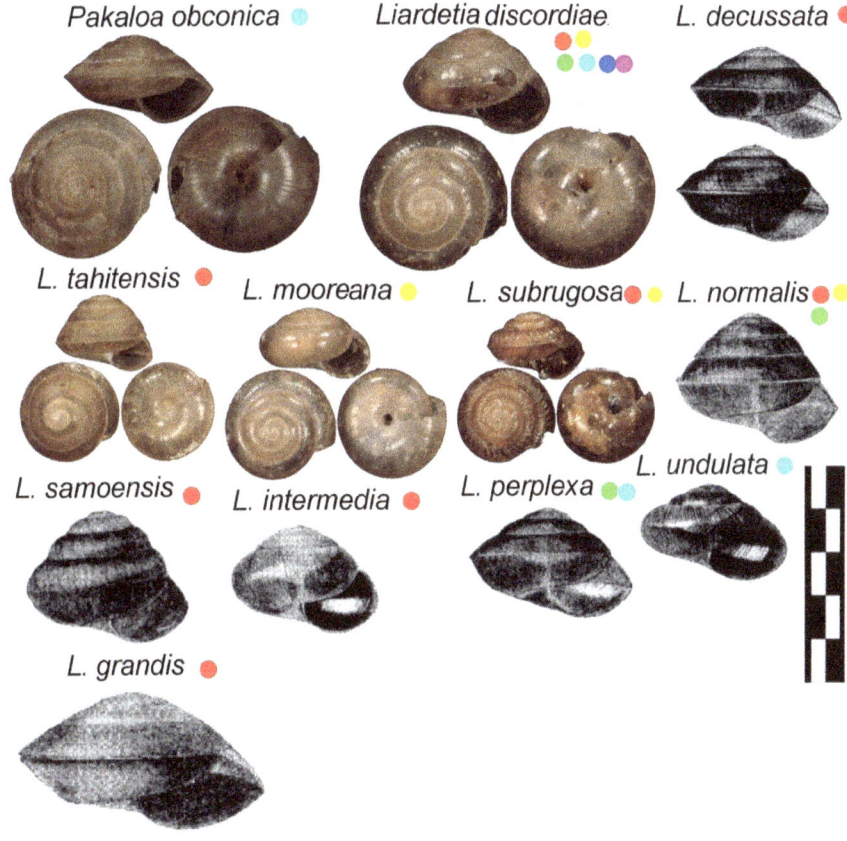

Liardetia (Liardetia) samoensis (Mousson, 1865)
   Liardetia (Liardetia) samoensis (Mousson) Solem 1959
   Liardetia (Liardetia) striolata (Pease) Baker 1938
   Pacific – Tahiti (Fautaua, on ground). Figure from Baker (1938).
Liardetia (Neoreus) grandis Baker, 1934
   Liardetia (Neoreus) grandis Baker, 1934
   Tahiti (north-west, above 700m, on vegetation). Figure from Baker (1938).
Liardetia (Ocenesia) discordiae (Garrett, 1881)
   Helix calculosa Gould, 1861
   Microcystis discordiae Garrett, 1881
   Liardetia (Ocenesia) discordiae (Garrett) Baker 1938
   Cooks, Australs - Mehetia, Tahiti, Moorea, Huahine, Raiatea), Tahaa, Bora Bora. Historically abundant under rotten wood and stones (Garrett 1884), now restricted to highest areas (below Mt. Tohiea on Moorea in 2010). Photos lectotype ANSP 49175.
Liardetia (Oceanesia) intermedia Baker, 1938
   Liardetia (Oceanesia) intermedia Baker, 1938
   Tahiti (north-west, above 1,600m, in moss on trees). Figure from Baker (1938).
Liardetia (Ocenesia) mooreana (Garrett, 1884)
   Zonites mooreana Garrett, 1884
   Liardetia mooreana (Garrett) Baker 1938
   Moorea (under wood and leaves). Photos lectotype ANSP 49159.
Pukaloa obconica (Pease, 1865)
   Helix obconica Pease, 1865
   Trochomorpha obconica (Pease) Pease 1871
   Trochonania obonica (Pease) Garrett 1884
   Nanina obconica (Pease) Tryon 1886
   Pukaloa obconica (Pease) Baker 1938
   Raiatea (recorded by Garrett from the upper parts of two valleys in the east and west). Photo syntype ANSP 49337.

Liardetia discordiae

Trochomorphidae
   Trochomorpha assimilis Garrett, 1884
      Trochomorpha assimilis Garrett, 1884
      Nanina Cressida (Gould) Binney 1884 misidentified
      Huahine (only recorded around Fare and Tefari.) Shells still present in 1991, now extinct. Photo lectotype ANSP 48964 and author's specimen.
   Trochomorpha cressida (Gould, 1846)
      Helix cressida Gould, 1846
      Trochomorpha cressida (Gould) Garrett 1884
      Helix vahine Hombron & Jacquinot, 1848
      Helix exclusa (Ferussac) Rousseau 1854
      Tahiti (Mt. Marau, Mt. Aorai and Fautaua, historically from above 250m, now only above 1,000m). In leaf litter.
   Trochomorpha pallens (Pease, 1871)
      Trochomorpha trochiformis pallens Pease, 1871
      Trochomorpha pallens (Pease) Garrett 1884
   Tahiti (Vaita in 1927, 100-300m), Moorea.
   Trochomorpha swainsoni (Pfeiffer, 1846)
      Helix swainsoni Pfeiffer, 1846
      Helix vahine 'Jaquinot' Pfeiffer 1849
      Helix depressiformis Pease, 1864
      Helix scuta 'Pease' Garrett 1884

Trochomorpha swainsoni   *T. typus*

Helix lenta 'Pease' Garrett 1884
Planamastra peasiana Hyatt & Pilsbry, 1911
Trochomorpha swainsoni (Pfeiffer) Pease 1871
Raiatea, on vegetation and on ground. Extinct. Records from Tahaa and Tahiti are errors. Photo NHMUK 'Tahiti' and author's specimen.

Trochomorpha typus Baker, 1841
Helix trochiformis Ferussac, 1821
Helix circumdata Anton, 1838
Trochomorpha trochiformis (Ferussac) Pease 1871
Trochomorpha typus Baker, 1941
Trochomorpha tahaensis Garrett label name (BPBM)
Raiatea, Tahaa. Extinct.

Superfamily Zonitoidea
Zonitidae
Hawaiia minuscula (Binney, 1840)
Hawaiia miniscula (Binney) Baker 1941
Pacific – Tahiti (1930s), Bora Bora. Photo NHMUK ('*boothiana* var.')

Superfamily Helicarionoidea
Helicarionidae
Ovachlamys fulgens (Gude, 1900)
Introduced - Tahiti, Moorea, Huahine, Raiatea, Tahaa, Bora Bora and Maupiti. First recorded on Tahiti and Raiatea in 1992.

Superfamily Limacoidea
Limacidae
  Deroceras laeve (Muller, 1774)
    Limax rarotonganus Heynemann Garrett 1884
    Introduced – Tahiti, Raiatea. First recorded on Tahiti in the 1800s.

Superfamily Arionoidea
Philomycidae
  An unidentified species present on Moorea since 2010 (FLMNH).

Systellommatophora
Superfamily Veronicelloidea
Veronicellidae
  Sarasinula plebeia (Fischer, 1868)
    Veronicella agassizi Cockerell, 1901
    Vaginula tahitiana Simroth, 1918
    Angustipes (Sarasinula) plebeius (Fisher) Solem 1964
    Sarasinula plebeia (Fisher) Gomes & Thome 2004
    Introduced (originally South American) – Tahiti, Raiatea. First recorded on Tahiti in 1901 and Raiatea in 1927 (as '*Veronicella* sp.'). Colour variable shades of brown with dark markings, no pale dorsal line.
  Laevicaulis alte (Férussac, 1822)
    Introduced – Moorea, Raiatea, Tahaa. First recorded on Moorea in 2010 (MNHN data) and Raiatea and Tahaa in 2017. Broader than *Sarasinula*, dark with pale line.
  Veronicella cubensis (Peiffer, 1840)
    Veronicella cubensis (Pfeiffer) Robinson & Hollingsworth 2009
    Introduced – Tahiti, Moorea, Raiatea. First identified in 2007. Colour brown with a light medial and two dark lateral bands, but these may be indistinct.

Deroceras laeve   *Sarasinula plebeia*   *Veronicella cubensis*

Philomycidae   *Laevicaulis alte*

References
Adams, H. & A. Adams 1858. *The genera of recent Mollusca; arranged according to their organization. In three volumes.* Van Voorest, London
Ancey, C.F. 1889. Description de mollusques nouveaux. *Le Naturaliste* 1889: 205
Anton, H.E. 1839. *Verzeichniss der Conchylien welche sich in der Sammlung von Herman Eduard Anton befinden.* Eduard Anton, Halle
Baker, H.B. 1938. Zonitid snails from Pacific Islands. Part 1. Southern genera of Microcystinae. *Bull. Bernice P. Bish. Mus.* 158: 1-102
Baker, H.B. 1940. Zonitid snails from Pacific islands—part 2. 2. Hawaiian genera of Microcystinae. *Bernice P. Bish. Mus. Bull.* 165: 105-20
Baker, H.B. 1941. Zonitid snails from Pacific Islands. Part 3 and 4. Genera other than Microcystinae and distribution and indexes. *Bull. Bernice P. Bish. Mus.* 166: 203-370
Baker, H.B. 1964. Type land snails in the Academy of Natural Sciences of Philadelphia. Part III, Limnophile and Thalassophile Pulmonata, Part IV, Land and freshwater Prosobranchia. *Proc. Acad. Nat. Sci. Phil.* 116: 148-193.
Bandel, K. 2001. The history of Theodoxus and *Neritina* connected with description and systematic evaluation of related Neritimorpha (Gastropoda). *Mitt. Geol.-Palaont. Inst. Univ. Hamburg.* **85**: 65-164
Binney, W.G. 1884. Notes on the jaw and lingual dentition of pulmonte mollusks. *Ann. N.Y. Acad. Sci.* 3: 79-136
Bland, T. & W.G. Binney 1872.On the lingual dentition of Nanina. *Am. Jour.Conch.* 7: 188-189
Boettger, O. 1880 Die *Pupa*-Arten Oceaniens. *Conch. Mitt.* 1(4): 45-72
Boettger, C.R. 1909. Die Molluskenausbeute der Hanseatischen Sudsee-Expedition 1909. *Abh. Senck. Naturf. Ges.* **353**
Bouchet P., J.-P. Rocroi, J. Frýda, B. Hausdorf, W. Ponder, Á. Valdés & A. Warén 2005. Classification and nomenclator of gastropod families. *Malacologia* 47(1-2): 1–397
Broderip, W.J. 1832. New species of shells collected by Mr. Cuming on the western coast of South America and in the islands of the South Pacific Ocean. *Proc. Comm. Sci. Corr. Zool. Soc. Lond.* **2**: 124-126
Broderip, W.J. 1833. Characters and descriptions of new species of Mollusca and Conchifera, collected by Mr. H. Cuming in 1827–1830. *Proc. Zool. Soc. Lond.* (1832): 194-202
Brot, A.L. 1877. Die Melaniaceen (Melanidae) in Abbildungen nach der Natur mit Beschreibungen. In: Kuster, H.C., ed., *Systematisches Conchylien-*

  Cabinet von Martini und Chemnitz. Neu herausgegeben und vervollstandigt. 1(24). Baur & Raspe, Nuremberg

Brook, F.J. 2010. Coastal landsnail fauna of Rarotonga, Cook Islands: systematics, diversity, biogeography, faunal history and environmental influences. *Tuhinga* **21**: 161-252

Carpenter, P.P. 1865 (for 1864). In Pease, W.H. 1865

Clench, W.J. & Turner, R.D. 1948. A catalogue of the family Truncatellidae with notes and descriptions of new species. *Occ. Pap. Moll.* 1(13): 157-212

Clessin, S. 1886 Forstsetzung on Kuster, H.C. & W. Dunker, in *Martini & Chemnitz Syst. Conch. Cab., Die Familie der Limnaeiden*

Cockerell, T.D.A. 1901. On a slug of the genus *Veronicella* from Tahiti. *Proc. U.S. Nat. Mus.* 23: 835–836

Cooke, C.M. & W.J. Clench. 1943. Land shells (Synceridae) from the southern and western Pacific. *Occ. Pap. Bernice P. Bish. Mus.* 17: 249–262

Cooke, C.M. & W.J. Clench. 1945. New Species of Succinea from Tahiti, with Remarks on other Polynesia species. *B.P. Bishop Museum, Occ. Papers* 18(8): 133-138

Cooke, C.M. Jr. & Kondo, Y. 1961. Revision of Tornatellinidae and Achatinellidae (Gastropoda, Pulmonata). *Bernice P. Bish. Mus. Bull.* 221[1960]: 1-303

Cowie, R.H 1998. Catalog of the nonmarine snails and slugs of the Samoan islands. *Br. Mus. Bull. Zool.* 3

Crosse, H. 1865. Addition a la note de M. le professeur A. Mousson sur la faune malacologique terrestre et fluviatile des archipels Viti et Samoa. *Journ. Conchyl.* 13(4): 430-31, pi. 14

Dohrn, H. 1859. Neue Landconchylien. *Malakozoologische Blätter* 6: 202-207

Dunker, R.W. in Dunker & Zelebor 1866. Bericht über die von der Novara-Expedition mitgebrachten Mollusken. *Verhandlungen der kaiserlich-königlichen zoologisch-botanischen Gesellschaft in Wien* 16: 909-916

Egorov, R. 2005. *Treasure of Russian Shells, Supplement 3, Part 1: A Review of the Genera of the Recent Terrestrial Pectinibranch Molluscs: Neritopsiformes: Hydrocenoidei, Helicinoidei.* Colus, Moscow

Férussac, A.E.J.P.J.F.D.A.D. 1821. *Tableaux systématiques des animaux mollusques classés en familles naturelles, dans lesquels on a établi la concordance de tous les systèmes; suivis d'un prodrome général pour tous les mollusques terrestres ou fluviatiles, vivants ou fossiles. Deuxième partie. (Première section.). Tableaux particuliers des mollusques terrestres et fluviatiles, présentant pour chaque famille les genres et espèces qui la composent. Classe des gastéropodes. Ordre des pulmonés sans opercules. II. Tableau systématique des Limaçons, Cochleae.* A. Bertrand, J.B. Sowerby, Paris, Londres. 90 pp.

Férussac, A.E.J.P.J.F. d'A. & G.P. Deshayes 1840. *Histoire naturelle générale et particulière des mollusques terrestres et fluviatiles tant des espèces que l'on trouve aujourd'hui vivantes, que des dépouilles fossiles de celles qui n'existent plus ; classés d'après les caractères essentiels que présentent ces animaux et leurs coquilles.* J.-B. Balliere, Paris

Fontanilla, I., Sta Maria, I., Garcia, J., Ghate, H., Naggs, F. & Wade, C. 2014. Restricted Genetic Variation in Populations of *Achatina* (Lissachatina) fulica outside of East Africa and the Indian Ocean Islands Points to the Indian Ocean Islands as the Earliest Known Common Source. *PLoS ONE* 9(9): e105151

Gargominy, O. 2008. Beyond the alien invasion: a recently discovered radiation of Nespopupinae (Gastropoda: Pulmonata: Vertiginidae) from the summits of Tahiti (Society Islands, Frenh Polynesia). *Journ. Conch.* 39(5): 517-536

Garrett, A. 1872. Descriptions of new species of land and fresh-water shells. *Amer. Journ. Conch.* 7(4): 219-3

Garrett, A. 1879. List of the land shells inhabiting Rurutu, one of the Austral Islands, with remarks on their synonymy, geographical range, and description of new species. *Proc. Acad. Nat. Sci. Phil.* 31: 17-30

Garrett, A. 1881.The terrestrial mollusca inhabiting Cook's or Hervey Islands. *Proc. Acad. Nat. Sci. Phil.* (2) 8(4): 381-411

Garrett, A. 1884. The terrestrial Mollusca inhabiting the Society Islands. *Journ. Acad. Nat. Sci. Phil. (II)* 9: 17-114.

Garrett, A. 1887. Mollusques terrestres des Iles Marquises (Polynésie). *Bull. Soc. Malacol. Fr.* 4: 1-48

Gerlach, J. 2016. *Icons of Evolution: tree-snails of the family Partulidae.* Phesuma Press, Cambridge

Gomes, S.R. & J.W. Thome 2004. Diversity and distribution of the Veronicellidae (Gastropoda: Soleolifera) in the Oriental and Australian biogeographic regions. *Mem. Queensland Mus.* 49(2): 589-601

Gould, A.A. 1846. Expedition Shells described for the work on the United State Exploring Expedition under the command of Charles Wilkes during the years 1838-1842. *Proc. Boston Soc. Nat. Hist.* 2: 142-152

Gould, A.A. 1848 (for 1847). [Untitled]. *Proc. Boston Soc. Nat. Hist.* **2**: 196-198

Gould, A.A. 1852. *Mollusca and shells, with an atlas of plates. United States Exploring Expedition: during the years 1838, 1839, 1840, 1841, 1842.*

Under the command of C. Wilkes, U.S.N. Vol. XII. Sherman, Philadelphia

Gould, A.A. 1859. Descriptions of new species of shells. *Proc. Boston Soc. Nat. Hist.* 7: 40-45

Gould, A.A. 1861. Description of new shells collected by the United States North Pacific Exploring Expedition. *Proc. Boston Soc. Nat.l Hist.* 7: 382–389; 400-409

Gray, J.A. 1825. A list and description of some species of shells not taken notice of by Lamarck. *Annals of Philosophy* (2)9: 407-415

Guillou, E. Le. 1841. Description de quatorze nerites nouvelles. *Rev. Zool. Soc. Cuvier.* 4(11): 343-4

Haynes, A. 2001. A revision of the genus *Septaria* Ferussac, 1803 (Gastropoda: Neritimorpha). *Ann. Naturhist. Mus. Wien* **103B**: 177-229

Hombron, J.B. & Jacquinot, H. 1842-1853. *Voyage au Pôle Sud et dans l'Océanie sur les corvettes l'Astrolabe et la Zélée pendant les années 1837-1838-1839-1840 sous le commandement de M. Dumont-d'Urville capitaine de vaisseau publié par ordre du gouvernement et sous la direction supérieure de M. Jacquinot, capitaine de Vaisseau, commandant de la Zélée. Zoologie. Atlas.* Gide & J. Baudry, Paris

Horst R. & M.M. Schepman. 1908. *Catalogue systématique des mollusques (gastropodes prosobranches et polyplacophores.* XIII. Leiden

Hyatt, A. & H.A. Pilsbry. 1911. *Manual of Conchology: structural and systematic.* (2). Pulmonata 21. Phuiladelphia

Johnson, R.I. 1964. The recent Mollusca of Augustus Addison Gould. *Bull. U.S. Natn. Mus.*

Kahn, J.G., C. Nickelsen, J. Stevenson, N. Porch, E. Dotte-Sarout, C.C Christensen, L. May, JS. Athens & P.V Kirch. 2014. Mid- to late Holocene landscape change and anthropogenic transformations on Moʻorea, Society Islands: A multi-proxy approach. *The Holocene* 25(2): 333-347

Kano Y. & T. Kase. 2003. Systematics of the *Neritilia rubida* complex (Gastropoda: Neritiliidae): three amphidromous species with overlapping distributions in the Indo-Pacific. *J. Moll. Stud.* 69(3): 273-284

Kondo, Y. 1962. The genus *Tubuaia* (Pulmonata, Achatinellidae). *Bernice P. Bish. Mus. Bull.* 224: 1–49

Lesson, R.P. 1831. *Voyage autourdu monde, execute par ordre du Roi, sur la corvette de sa Majeste, La Coquille, pendant les annees 1822, 1823, 1824 et 1825, sous le ministere et conformement aux instructions de*

S.E.M. le Marquis de Clermont-Tonnerre, Ministre de la Marine; et publie sous les auspices de son Excellence Mgr. Le Cte De Chabrol, Ministre de la Marine et des Colonies. Histoire naturelle. Zoologie..2(1). Bertrand, Paris

Martesn, E. von 1860. Die Heliceen nach natiirlicher Verwandtschaft systematisch geordnet von Joh. Christ. Albers. Zweite Ausgabe. W. Engelmann, Leipzig

Martens, E.von 1866. XXIII. Conchological Gleanings. *Ann. Mag. Nat. Hist.* **17**: 99, 202-213

Martens, E. von 1879. Die Gattung Neritina. In: Kiister, H.C., ed., *Systematisches Conchylien-Cabinet von Martini und Chemnitz. Neu herausgegeben und vervollstandigt.* 2(10). Baur & Raspe, Nurenberg

Martens, E. von & B. Langkavel. 1871. Eine Sammlung von Südsee-Conchylien. *Donum Bismarkianum.* F. Berggold, Berlin

Meyer, W.M., N.W. Yeung, J. Slapcinsky & K.A. Hayes. 2017. Two for one: inadvertent introduction of *Euglandina* species during failed bio-control efforts in Hawaii. *Biological Invasions* 19(5): 1399

Mousson, A. 1869. Faune malacologique terrestre et fluviatile des iles Samoa, publiée d'après les envois de M. le Dr E. Graeffe. *Journ. Conchyl.* 17(4): 323-390

Paetel, F. 1873. *Catalog der Conchylien-Sammlung: nebst Uebersicht des angewandten Systems.* Gebruder Paetel, Berlin

Patterson, C.M. 1989. Morphological studies of a Tahitia succineid, *Succinea wallisi. Malacol. Rev.* 22: 17-24

Pease, W.H. 1861. Descriptions of New Species of Mollusca from the Pacific Islands. *Proc. Zool. Soc. Lond.* 29: 242-247

Pease, W.H. 1865 (for 1864). Descriptions of New Species of Land Shells from the Islands of the Central Pacific. *Proc. Zool. Soc. Lond.* **32**: 668-676

Pease, W.H. 1865. Descriptions of New Species of Land Shells from the Islands of the Central Pacific. *Proc. Zool. Soc. Lond.* 32(1864): 668-676

Pease, W.H. 1867. Illustrations of new species of *Partula. Amer. Journ. Conch.* 3(1): 81

Pease, W.H. 1868. Description of a new genus and eleven species of land shells, inhabiting Polynesia. *Amer. Journ. Conch.* 4(3): 153-160

Pease, W.H. 1869. Monographic de la famille des Realiea, Pfeiffer. *Journ. Conchyl.* 17: 131-60.

Pease, W.H. 1870. Observations sur les especes de coquilles terrestres qui habitent l'fle de Kauai (Ties Hawaii), accompagn6es de descriptions d'especes nouvelles. *Journ. Conchyl.* 18: 87-97

Pease, W.H. 1871. Remarques sur certaines espèces de coquilles terrestres, habitant la Polynésie, et description d'espèces nouvelles. *Journ. Conchyl.* 18(1870) (4): 393-403

Pfeiffer, L. 1849 (for 1848). Descriptions of twenty-four new species of Helicea, from the collection of H. Cuming, Esq. *Proc. Zool. Soc. Lond.* **16**: 126-131

Pfeiffer, L. 1850. Beschreibungen neuer Landschnecken. *Zeitsch. Malakozool.* 7(6): 65-80

Pfeiffer, L. 1851. Conspectus cyclostomaceorum (contin.). *Zeitsch. Malakozool.* 8(11): 161-76

Pfeiffer, L. 1852. *Monographia pneumonopomorum viventium. Sistens descriptiones systematicas et criticas omnium hujus ordinis generum et specierum hodie cognitarum, accedente fossilium enumeratione.* Cassellis (Th. Fischer), London, Paris

Pfeiffer, L. 1853. Diagnosen neuer Heliceen. *Zeitschr. Malak.* 10(4): 51-58

Pfeiffer, L. 1854. A Monograph of the Genera Realia and Hydrocena. *J. Zool.* 22(1): 304-309

Pfeiffer, L. 1855. Descriptions of forty-two new species of Helix, from the collection of H. Cuming, Esq. *Proc. Zool. Soc. Lond.* 22[1854]: 49-57

Pfeiffer, L. 1858. *Monographia pneumonopomorum viventium. Sistens descriptiones systematicas et criticas omnium hujus ordinis generum et specierum hodie cognitarum, accedente fossilium enumeratione.* Suppl. 1. Cassellis (Th. Fischer), London, Paris

Pfeiffer, L. 1859. *Monographia heliceorum viventium, sistens descriptions systematicas et criticas omnium huis familiae generum et specierum hodie cognitarum.* F.A. Brockhaus, Leipzig. **4**

Pfeiffer, 1862(for 1861). Description of sixteen new species of land-shell from the collection of H. Cuming, Esq. *Proc. Zool. Soc Lond.* **29**: 386-389

Pfeiffer, L. 1868. *Monographia heliceorum viventium. Sistens descriptiones systematicas et criticas omnium huius familiae generum et specierum hodie cognitarum. Volumen sextum. Supplementum tertium. Sistens enumerationem auctam omnium huius familiae generum et specierum hodie cognitarum. Accedentibus descriptionibus novarum specierum.* 2. F.A. Brockhaus, Leipzig

Pfeiffer, L.G.K. 1876. *Monographia heliceorum viventium, sistens descriptiones systematicas et criticas omnium hujus familiae generum et specierum hodie cognitarum.* 7. F.A. Brockhaus

Pilsbry, H.A. 1893. Manual of conchology: structural and systematic. With illustrations of the species. By George W. Tryon, Jr. Second series:

Pulmonata. Vol. IX. (Helicidae, vol. 7.). *Guide to the study of helices.* Academy of Natural Sciences, Philadelphia

Pilsbrty, H.A. 1906-7. *Manual of conchology. Structural and systematic. With illustrations of the species. Founded by George W. Tryon, Jr. Second series: Pulmonata. Vol. XVIII. Achatinidae: Stenogyrinae and Coeliaxinae.* Academy of Natural Sciences, Philadelphia

Pilsbry, H.A. 1916. *Manual of conchology. Second series: Pulmonata. Vol. XXIV. Pupillidae (Gastrocoptinae).* Academy of Natural Sciences, Philadelphia

Pilsbry, H.A. 1918-20. *Manual of conchology. Second series: Pulmonata. Vol. XXV.* Academy of Natural Sciences, Philadelphia

Pilsbry, H.A. & Cooke, C.M., Jr. 1915. *Manual of conchology. Second series: Pulmonata. Vol. XXIII. Appendix to Amastridae. Tomatellinidae. Index, vols. XXI-XXIII.* Academy of Natural Sciences, Philadelphia

Preece, R.C. 1995. Systematic review of the landsnails of the Pitcairn Islands. *Biol. Journ. Linn. Soc.y* 56: 273–307

Recluz, C.A. 1850. Notice sur le genre nerita et sur le S.-G. Neritina, avec le catalogue synonymique des neritines. *Journ. Conchyl.* 1(2): 131-64

Recluz, C.A. 1852. *Jour. de Conch.* 3: 413

Reeve, A.L. 1852 *Conchologia Iconica: or, Illustrations of the Shells of Molluscous Animals. Volume VII. Containing a monograph of the genus Helix. L..* Reeve, London

Reeve, A.L. 1860. *Conchologia Iconica: or, illustrations of the shells of molluscous animals. Volume XII. Containing monographs of the genera Argonauta. Nautilus. Terebra. Aspergillum. Thracia. Melania. Hemisinus. Anculotus. Metatoma. Io. Pirena. Melanopsis. Scarabus. Trigonia. Myochama.* L. Reeve, London.

Riech, E. 1937. Systematische, anatomische, okologische und tiergeographische Untersuchungen iiber die SiiBwasser-mollusken Papuasiens und Melanesiens. *Archivfur Naturgeschichte* (N.F.) 6(1): 37-153

Resch, V.H., J.R. Barnes & D.A. Craig 1990. Distribution and ecology of benthic macrinvertebrates in the Opunohu river catchment, Moorea, French Polyensia. *Annls. Limnol.* 26(2-3): 195-214

Richling I. & P. Bouchet P. 2013. Extinct even before scientific recognition: a remarkable radiation of helicinid snails (Helicinidae) on the Gambier Islands, French Polynesia. *Biodiversity and Conservation.* 22: 2433-2468

Robinson, D.G.& R.G. Hollingsworth. 2004. Survey of slug and snail pests on subsistence and garden crops in the islands of the American Pacific:

Guam, and the Northern Mariana Islands: Part I. The Leatherleaf Slugs (Family: Veronicellidae). 1-11. United States Department of Agriculture, Animal and Plant Health Inspection Service.

Rousseau, L. 1854. *Description des mollusques, coquilles et Zoophytes. In: Voyage au Pole Sud et dans l'Oceanie sur les corvettes l'Astrolabe et la Zelee; execute par ordre du roi pendant les annees 1837-1838-1839-1840, sous le commandement de M. J. Dumont-d'Urville, capitaine de vaisseau; publie par ordre du gouvernement, sous la direction superieure de M. Jacquinot, capitaine de Vaisseau, commandant de la Zelee. Zoologie par Mm. Hombron etJaquinot.* 5. G. & J. Baudry, Paris

Schmeltz, J.D.E. 1874. *Museum Godeffroy Catalog V Nebst einer Beilage enthaltend topographische und zoologische Notizen.* L. Friedrichsen & Co., Hamburg

Schmeltz, J.D.E. 1877. *Museum Godeffroy Catalog VI. Nachrage zu Catalog V.* L. Friedrichsen & Co., Hamburg

Shuttleworth, R.J. 1852. Diagnosen neuer Mollusken. *Mitt. Naturf. Gesell. Bern* (260/261) 54: 289-304

Simroth, H. 1918. Uber einige Nacktschnecken vom malayischen Archipel von Lombok an ostwarts bis zu den Gesellschafts-Inseln. *Abh. Heraus. Senckenb. Naturf. Gesell.* 35(3): 259-302

Solem A, 1964. New records of New Caledonian non-marine mollusks and an analysis of the introduced mollusks. *Pac. Sci.* **18**: 130-137

Solem, A. 1976. *Endodontoid Land Snails from Pacific Islands (Mollusca: Pulmonata: Sigmurethra). Part 1. Family Endodontidae.* Field Museum Press, Chicago

Solem, A. 1982. *Endodontoid land snails from Pacific islands (Mollsuca: Pulmonata: Sigmurethra). Part 1.* Field Museum of Natural History, Chicago.

Solem, A. 1983. *Endodontoid land snails from Pacific islands (Mollusca: Pulmonata: Sigmurethra). Pan II. Families Punctidae and Charopidae. Zoogeography.* Field Museum of Natural History, Chicago

Sowerby G.B. 1833. [Descriptions of new species of shells from the collection formed by Mr. Cuming on the western coast of South America, and among the islands of the southern Pacific Ocean.]. *Proceedings of the Committee of Science and Correspondence of the Zoological Society of London*, 2 (1832): 194-202

Sowerby, G.B. in Gray, J. E. 1839. Molluscous animals, and their shells. *in:* Richardson, J., Vigors, N.A., Lay, G.T., Bennett, E.T., Owen, R., Gray, J.E., Buckland, W. & Sowerby, G.B. [Eds]. *The zoology of Captain*

Beechey's voyage; compiled from the collections and notes made by Captain Beechey, the officers and naturalist of the expedition, during a voyage to the Pacific and Behring's Straits performed in His Majesty's ship Blossom, under the command of Captain F.W. Beechey, R.N., F.R.S. &c. &c. in the years 1825, 26, 27, and 28. Henry G. Bohn, London: 101-155

Sowerby G.B., II. 1842. Monograph of the Genus Helicina. *in: Thesaurus conchyliorum, or Monographs of genera of shells*. Sowerby, London

Stàrmiilhner, F. 1976. Beitrage zur Kenntnis der SusswasserGastropoden pazifischer Inseln. Ergebnisse der Osterreichischen Indopazifik-Expedition 1971 des 1. Zoologischen Institutes der Universitât Wien. *Ann. Naturg. Mus. Wien* 80: 473-676

Starmuhlner, F. 1984. Checklist of the fauna of mountain streams of tropical Indopacific Islands. *Ann. Naturhist. Mus. Wein* 88/89(B): 457-480

Tilier, S. & B.C. Clarke 1983. Lutte biologique et destruction du patrimoine génétique: le cas des mollusques gastéropodes pulmonés dans les territories francais du Pacifique. *Gen. sel. Evol.* 15:559

Tryon, G.W. 1866. Notes on Mr Pease's species of Polynesian Phaneropnemona. *Amer. Journ. Conchol.* 1: 82

Tryon, G.W. 1886. *Manual of conchology. Second series: Pulmonata. Vol. II. Zonitidae.* Academy of Natural Sciences, Philadelphia. 265 pp

Tryon, G.W. 1887. *Manual of conchology. Vol. III.* Academy of Natural Sciences of Philadelphia

Wagner, A.J. 1905. Helicinenstudien. *Denksch. Math.-Naturw. Klasse Kaiser. Akad. Wiss.* 77: 357-450

Wagner, A.J. 1907-1911. Die Familie der Helicinidae. Neue Folge. *in*: Küster, H.C. & Kobelt, W. [Eds]. *Systematisches Conchylien-Cabinet von Martini und Chemnitz.* Nürnberg. 1 (18 (2)): 1-391 + pl. 1-70.

Walker, J.C., 1988. Classification of Australian buliniform planorbids (Mollusca: Pulmonata). *Records of the Australian Museum* 40(2): 61–89

www.ingramcontent.com/pod-product-compliance
Lightning Source LLC
Chambersburg PA
CBHW040518220526
45473CB00012B/2903